T0134672

Introduction to *Mathematica*® with Applications

Marian Mureşan

Introduction to *Mathematica*® with Applications

 Springer

Marian Mureşan
Faculty of Mathematics and Computer Science
Babeş-Bolyai University
Cluj-Napoca, Romania

ISBN 978-3-319-84794-8   ISBN 978-3-319-52003-2   (eBook)
DOI 10.1007/978-3-319-52003-2

Printed on acid-free paper

This Springer imprint is published by Springer Nature
The registered company is Springer International Publishing AG
The registered company address is: Gewerbestrasse 11, 6330 Cham, Switzerland

To my wife Viorica, always with love

Foreword

The book (written by Professor Marian Mureşan) is intended to present the *Mathematica* system in a manner in which the reader will find it easy to acquire a part of the considerable number of mathematical instruments offered by the producer, based on a large variety of examples taken from different scientific branches.

The importance of the symbolic calculus is incontestable, since nowadays the scientific calculus is not reduced to the numerical one. Modern mathematical problems, or those appearing through modeling natural phenomena, lead to such complicated symbolic expressions that we need the help of computers to process them. The fast development of the computers' technical capacity has enabled the genesis of a new domain—the symbolic (or formal) calculus—through which a genuine human–machine collaboration is achieved. The symbolic calculus systems are aimed at automating difficult calculi. They provide simple access to many sophisticated mathematical instruments, using a simple yet flexible language, that allows avoidance of writing interminable subroutines (similarly to how it happens with utilizing traditional languages).

It is mainly mathematical aspects that are covered in this book, so that the reader can understand as precisely as possible how to handle the problem. At the same time, the user is warned with regard to the type of problems which can be solved by means of the computer and the necessity of the analysis and verification of the results. Special emphasis is given to the fact that a session of *Mathematica* is typically interactive. Often there are multiple ways to approach problems, the choice of one of them being made on the spot, according to the answer of the system. The reader is carefully guided to the most suitable choice.

The book contains 11 chapters on specific aspects of the *Mathematica* language or branches of mathematics in general, such as ordinary differential equations, to which Chap. 7 is dedicated. It is worth mentioning that part of the applications presented in the book have been elaborated by the author in recent years, being the subject of certain scientific articles mentioned in the bibliography. In this book, standard linguistic constructions are used.

The book is at an incredibly high scientific level, and it is useful to many categories of users; for researchers, professors, and students in scientific fields, the knowledge of such a symbolic calculus system is nowadays indispensable. Mathematics researchers can use the programs for testing some conjectures or even for proving some results that need a big amount of calculation (in differential geometry, number theory, combinatorics, theory of functions of complex variable, numeric calculus, differential equations). The book is useful for engineers, economists, and IT specialists, who can therefore benefit from easy access to a volume of vast mathematical knowledge. I would mention that parts of the content of this book have been presented during the scientific seminars held at the Faculty of Mathematics and Computer Science, having been highly appreciated. Thanks to all of the above mentioned, I strongly recommend Prof. Marian Mureşan's book.

Cluj-Napoca, Romania   Prof. Valeriu Anisiu
August 2016

Preface

This book is the output of our work over several years. Being involved in calculus of variations and optimal control problems, we have realized that an exact calculation and a suggestive visualization are very useful, making the ideas addressed many times only in an ε-δ language clearer. Then we have chosen *Mathematica* for computations and visualizations of the ideas. Why did our option go to *Mathematica*? The answer is simple: because we had noticed the wonderful results of Prof. J. Borwein and his colleagues regarding the decimals of number π. Their approach was based on an extensive use of *Mathematica*.

Wolfram Research, located at Champaign, IL, USA, is the company which has been developing *Mathematica*.

Mathematica is continuously developing. We used *Mathematica* 10.3. It is very likely there will be newer versions with extra facilities in the future.

We have introduced notions and results in *Mathematica* in our lectures to master students at the Faculty of Mathematics and Computer Science of the Babeş-Bolyai University in Cluj-Napoca, Romania. We did the same thing with our PhD students at three summer schools organized in the framework of the grant "Center of Excellence for Applications of Mathematics" supported by DAAD, Germany. The summer schools have been organized in Struga (Macedonia, FYROM), Sarajevo (Bosnia and Herzegovina), and Cluj-Napoca (Romania).

This book is not very large, but it collects many examples. In the first part of the book, the examples are discussed in detail helping the reader to understand the reasoning in and with *Mathematica*. Later on, the reader is led to use the benefit of the Help and other sources freely offered by Wolfram Research. We take into account mainly the Wolfram community forum as well as the video training and conferences generously offered by Wolfram Research.

A well-motivated case for visualization in mathematics is contained in [58].

Here is the right place to express my gratitude to the following colleagues of mine from the Faculty of Mathematics and Computer Science of the Babeş-Bolyai University for their support: Anca Andreica, Valeriu Anisiu, Paul Blaga, Virginia

Niculescu, Adrian Petruşel, and Adrian Sterca. The existence and development of the MOS (Modeling, Optimization, and Simulation) Research Center of our faculty was a real help for us in the preparation of this book.

Cluj-Napoca, Romania   Prof. Marian Mureşan
August 2016

Contents

Chapter 1
About *Mathematica*

1.1 Introduction

Mathematica is a product of Wolfram Research on the address http://www.wolfram.com/. The complete documentation for an effective use of *Mathematica* can be found at that page. The documentation is clear in explanation, rich in examples, and fundamental for everybody working with *Mathematica*. Therefore, we avoided repeating explanations of the rules of arithmetic operations, syntaxes of the commands, built-in functions, etc. We let this pleasure to the reader to discover them by the references given herein.

The evolution of the logos of *Mathematica* is introduced at http://mathworld.wolfram.com/Spikey.html and presented below by Fig. 1.1.

The versions of spikeys are shown in Fig. 1.2

```
GraphicsGrid[Partition[
MapIndexed[Show[#1,PlotLabel→"Version "<>ToString[#2[[1]]]] &,
Take[spikeys,10]],4,4,{1,1},{}],ImageSize→550,Alignment→Top]
```

and all are trademarks of Wolfram Research.

Along the present book, we used *Mathematica* 10.3; its logo is given in Fig. 1.3.

Mathematica is a very versatile and powerful package for numeric, symbolic, and graphic calculus and even much more. Besides the documentation for *Mathematica*, there are a lot of interesting books leading the reader into this miraculous world. We mention here only a few titles [11, 20, 32, 33, 64–68, 72, 73], and [16]. The documentation issued by Wolfram Research is of main importance [75] and [74]. A very important and useful activity is the series of video workshops generously offered by Wolfram Research.

© Springer International Publishing AG 2017
M. Mureşan, *Introduction to Mathematica*® *with Applications*,
DOI 10.1007/978-3-319-52003-2_1

`spikeys = {`

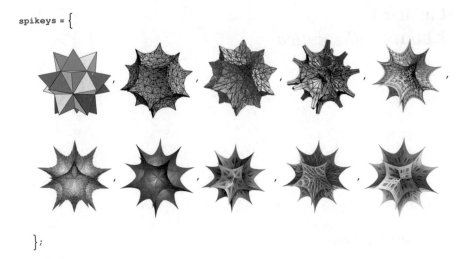

`};`

Fig. 1.1 Spikeys

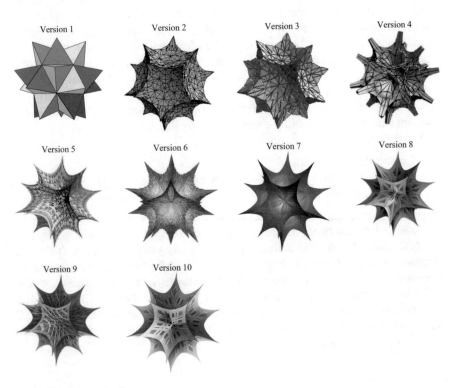

Fig. 1.2 Versions of spikeys

Fig. 1.3 The spikey of this
version

1.1.1 Warning

This book is intended to be only an introduction to *Mathematica*, mainly from the point of view of a mathematician according to his experience. *Mathematica* is much larger than we present it here and is continuously growing fast. Therefore, it is hard to predict how it will look like over a decade.

Chapter 2
First Steps to *Mathematica*

2.1 The Introductory Techniques for Using *Mathematica*

When we intend to write a document in *Mathematica,* we firstly select from Format menu the StyleSheet option and choose the type of document we are interested in. Then we follow clicking either Text or Input in Style of Format, depending on whether we want to introduce text or formulas to compute (evaluate). For various formulas, the option Writing Assistant in Palettes is of real help.

2.1.1 Numbers

Suppose we want to compute the sum $2 + 5$. We type

    2+5

Then immediately appears

    2+5   (* Here we press SHIFT+ENTER to get the result *)
    7

Thus the result is 7. As one can already see, we have inserted a comment. The comments are useful when we introduce initial data and/or mark different parts of a longer code. They are written in gray. Below we introduce other examples with arithmetic operations.

    25-13   (* 25 minus 13 SHIFT+ENTER to get the result *)
    12

    22*11   (* 22 times 11 SHIFT+ENTER to get the result *)
    242

    22/11   (* 22 divided by 11 SHIFT+ENTER to get the result *)
    2

© Springer International Publishing AG 2017
M. Mureşan, *Introduction to Mathematica® with Applications,*
DOI 10.1007/978-3-319-52003-2_2

22^11   (* or *) 22^{11} (* 22 at the power 11 SHIFT+ENTER to get the result *)
584318301411328

7.1/2.33   (* Division *)
3.04721

7/6   (* Division *)
$\frac{7}{6}$ (* The result is given as a fraction. If we wish a result under the form of a decimal number, use the built-in function N[ ] and write *)

N[7/6]   (* The standard output is with 5 decimals. Here we use the built-in function N; the arguments of a function, built-in functions included, are given by means of square brackets *)
1.16667

We can get the same result writing

7/6//N   (* This means that the built-in function N[ ] is applied to the fraction 7/6 *)
1.16667

Also,

7./6   (* We write either term as a decimal number and the result is returned as decimal number *)
1.16667

N[7/6,10]   (* If we need more than 5 decimals, we specify their number in the second argument of the built-in function N[ ] *)
1.166666667

We now mention another format for the built-in function N

N@7/6   (* This means the built-in function N[ ] is applied to the fraction 7/6; equivalently, we write N[7/6] *)
1.16667

Suppose we want to know the first 15 decimals of the number pi, denoted Pi or π. If we write either of the inputs

Pi   (* or *) π (* Input *)

we get that

π (* and *)
π

i.e., no numerical result is returned. To get the expected numerical result, we have to write

N[Pi,15]
3.14159265358979

or

N[π,15]

3.14159265358979

and *Mathematica* returns it with 15 decimals. Another example is given below:

$\left(\frac{1}{2}\right)^8$ (* We wrote this expression with the Writing Assistent menu in Palletes *)

$\frac{1}{256}$

If we want a numerical result with 10 decimals, we write

N$\left[\left(\frac{1}{2}\right)^8, 10\right]$

0.003906250000

Two interesting cases follow:

2/0
2/∞
Power::infy: Infinite expression $\frac{1}{0}$ encountered. »
ComplexInfinity
0

In the first case, the answer contains a warning message and a result (ComplexInfinity), while in the second case the result is correct according to the usual convention [53, p. 30].

We now present some notations and built-in functions.

$\mathbb{N}, \mathbb{Z}, \mathbb{R}, \mathbb{C}$ (* Used for the sets of natural, integer, real, and complex numbers *)

x+y I   (* The complex number x+iy; I is reserved for the imaginary unit $\sqrt{-1}$ *)

Re[z]   (* The real part of a complex number z; Re is a built-in function *)

Im[z]   (* The imaginary part of a complex number z; Im is a built-in function *)

Conjugate[z]   (* The complex conjugate of a complex number z; Conjugate is a built-in function *)

Abs[z]   (* The absolute value of a complex number z; Abs is a built-in function *)

Arg[z]   (* The argument θ in $z = |z|(\cos\theta + i\sin\theta)$, the trigonometric form of a complex number z; Arg is a built-in function *)

$\mathbb{N}, \mathbb{Z}, \mathbb{R}, \mathbb{C}$
x+y I
Re[z]
Im[z]
Conjugate[z]
Abs[z]
Arg[z]

Suppose we want to see whether $\sqrt{3} + \sqrt{5} + \sqrt{7} + \sqrt{11}$ is an algebraic number. Using the Writing Assistant Palette, we write $\sqrt{3} + \sqrt{5} + \sqrt{7} + \sqrt{11} \in$ Algebraics and understand it as a question, i.e., "is it true that the sum belongs to the set of algebraic numbers?"

$\sqrt{3} + \sqrt{5} + \sqrt{7} + \sqrt{11} \in$ Algebraics
True

Another way to check that the sum is an algebraic number consists in performing the following steps. First we generate the list $\{3, 5, 7, 11\}$, transform it as $\{\sqrt{3}, \sqrt{5}, \sqrt{7}, \sqrt{11}\}$, add its elements, and finally we verify whether it is true that their sum is an algebraic number or not.

Plus@@Sqrt[{3,5,7,11}]∈ Algebraics   (* First we mention that Sqrt[{3,5,7,11}] is equivalent to $\left\{\sqrt{3}, \sqrt{5}, \sqrt{7}, \sqrt{11}\right\}$. Here @@ is a substitution for the built-in function Apply and the left-hand side is interpreted Apply$\left[$Plus, $\left\{\sqrt{3}, \sqrt{5}, \sqrt{7}, \sqrt{11}\right\}\right]$, which gives $\sqrt{3} + \sqrt{5} + \sqrt{7} + \sqrt{11}$. Finally we ask if this sum is an algebraic number or not *)
True

2.1.1.1 Divisibility

We want to find out how many zero digits the number 32! has at the end.

IntegerQ$\left[32! / 10^7\right]$ (* Is 32!/10^7 an integer? Yes. This means that 32! ends with at least 7 zeros *)
IntegerQ$\left[32! / 10^8\right]$ (* Is 32!/10^8 an integer? No. This means that 32! ends with exactly 7 zeros *)
True
False

Another example of divisibility follows below.

IntegerQ[75^75 * 98^98/10^98]   (* Is 75^75 * 98^98/10^98 an integer? The answer is yes *)
Mod[75^75 * 98^98, 10^98]   (* We test the same thing with the built-in function Mod *)
IntegerQ[75^75 * 98^98/10^99]   (* The answer is no and we conclude the number 75^75 * 98^98 has exactly 98 zero digits at the end *)
True
0
False

2.1.2 *Bracketing in Mathematica*

There are four kinds of bracketing in *Mathematica*.

(term)	parentheses for grouping
f[x]	square brackets for arguments of functions
{a,b,c,d}	curly braces for lists. A list is a sequence of elements in which the order counts; it is not a set
a[[i]]	the ith element of the list a.

Other examples of calculations by similar constructions are introduced below together with the corresponding results.

2.1.3 *Set or SetDelayed Operator*

The = operator (called the Set operator) evaluates the expression on its right-hand side only once, when the assignment is made.

x=3   (* or *)   Set[x,3]

The value of the variable x is 3 and it is permanent. It is kept until a different assignment regarding it occurs or is explicitly removed.

A certain value can be simultaneously set to several variables as

x=y=value   (* The same value is set to both variables *)
value

To remove the value of a variable, we write

x=.   (* or *)   Clear[x]

The := operator (called the SetDelayed operator) used in defining functions differs from the = operator. When we use := the expression appearing on its right is evaluated anew each time the expression appearing on its left is called. There are situations when = and := can be used interchangeably. There are cases when one is appropriate and the other is not.

a:=3   (* or *)   SetDelayed[a,3]

Mathematica uses both upper- and lowercase letters. There is a convention in *Mathematica* that the built-in objects always have names starting with uppercase (capital) letters. To avoid confusion, one should always choose names for one's own variables that start with lowercase letters.

mnop	is a variable (independent or dependent) name introduced by the user
Mnop	is a built-in object whose name begins with a capital letter

Here is a short list of some built-in functions.

Divide[x, y]	(* x/y *)
Sqrt[x]	(* $\sqrt{x}$ *)
Log[x]	(* $\log_e x$ *)
Sin[x]	(* $\sin x$ *)
Tan[x]	(* $\tan x$ *)
ArcSin[x]	(* $\arcsin x$ *)
ArcTan[x]	(* $\arctan x$ *)
Tanh[x]	(* $\tanh x$ *)
ArcTanh[x]	(* $\operatorname{arctanh} x$ *)

We note the following facts regarding the variables:

x y	means x times y
xy	is the variable xy. We note that there is no space between x and y
5x	means 5 times x
x^2 y	means $x\hat{\ }2$ times y or x^2 times y

The following shortcuts are useful:

@	means that $f @ expr$ and is defined $f[expr]$
@@	means Apply. $f @@ \{a, b, c, d\} = f[a, b, c, d]$
@@@	means applying at level 1. $f @@@ \{\{a,b,c\},\{d,e\}\}$ is $\{f[a,b,c],f[d, e]\}$
/.	means ReplaceAll. $\{x, y\}/. x \to a$ is $\{a, y\}$
&	specifies a pure function. $\#1^3 + \#2^4$ &[x,y] is $x^3 + y^4$
&&	is the logical And

2.1.4 Some Simple Steps

Clear[a,b,c,x]   (* Clear is a very useful built-in function. It removes any previous definition regarding the arguments of it *)
{Plus@@{a,b,c},   (* Here @@ is again a substitution for the built-in function Apply and the code is interpreted as Apply[Plus,{a,b,c}] which gives a + b + c. a, b, and c are any objects that can be added: complex numbers, real numbers, matrices that have the same number of rows and the same number of columns, etc. To avoid long lines of results, we put all these commands in one list *)

Plus@@(a × b × c),

Plus@@(a/b/c),

Plus@@{a × b × c}   (* This is Apply[Plus, {a × b × c}], in the curly braces we have precisely one number *)}

Power[x, y]==Power@@{x, y}== Power@@(x + y)==

Power@@(x × y)==x^y (* We conclude that we can write x^y in either way *)

$$\left\{ a + b + c, a + b + c, a + \frac{1}{b} + \frac{1}{c}, abc \right\}$$

True

Range@10   (* The list of natural numbers from 1 to 10, i.e., 1,2,3,4,5,6,7,8,9,10 *)

Table[Range@n,{n,10}]   (* Table produces a list, in this case is a list of lists, i.e., {Range[1], Range[2], Range[3], Range[4], Range[5], Range[6], Range[7], Range[8], Range[9], Range[10]} *)

Flatten[Table[Range@n,{n,10}]]   (* From the previous list of lists it returns one list deleting all internal curly braces. Thus we get a single list *)

Plus@@Flatten[Table[Range@n,{n,10}]]   (* The elements of the previous list are added *)

{1,2,3,4,5,6,7,8,9,10}

{{1},{1,2},{1,2,3},{1,2,3,4},{1,2,3,4,5},{1,2,3,4,5,6},{1,2,3,4,5,6,7}, {1,2,3,4,5,6,7,8},{1,2,3,4,5,6,7,8,9},{1,2,3,4,5,6,7,8,9,10}}

{1,1,2,1,2,3,1,2,3,4,1,2,3,4,5,1,2,3,4,5,6,1,2,3,4,5,6,7,1,2,3,4,5,6,7,8, 1,2,3,4,5,6,7,8,9,1,2,3,4,5,6,7,8,9,10}

220

The last three commands can be written under a shorter form.

Table[Range@n,{n,10}]

Flatten@%   (* % takes the last result *)

Plus@@%

{{1},{1,2},{1,2,3},{1,2,3,4},{1,2,3,4,5},{1,2,3,4,5,6},{1,2,3,4,5,6,7}, {1,2,3,4,5,6,7,8},{1,2,3,4,5,6,7,8,9},{1,2,3,4,5,6,7,8,9,10}}

{1,1,2,1,2,3,1,2,3,4,1,2,3,4,5,1,2,3,4,5,6,1,2,3,4,5,6,7,1,2,3,4,5,6,7,8, 1,2,3,4,5,6,7,8,9,1,2,3,4,5,6,7,8,9,10}

220

We can show other examples.

Range@10;   (* The semicolon ";" stops printing the list {1, 2, 3, 4, 5, 6, 7, 8, 9, 10} *)

Table[Range@n,{n,10,10}]   (* This is a list containing only the list {1,2,3,4,5,6,7,8,9,10} *)

Plus@@Table[Range@n,{n,10,10}]   (* We get the list {1, 2, 3, 4, 5, 6, 7, 8, 9, 10} *)

{Plus@@Plus@@Table[Range@n,{n,10,10}]   (* Now we add the
elements of the list *)
Plus@@Range@10   (* This addition can be performed easier in
this way *)}
Times@@Times@@Table[Range@n,{n,10,10}]==
Times@@Range[10]==10!
{1,2,3,4,5,6,7,8,9,10}
{1,2,3,4,5,6,7,8,9,10}
{55,55}
True

Some example with means follow.

Mean@@Table[Range@n,{n,10,10}]==Mean@@{Range@10}==
Plus@@Range@10/10==11/2   (* Here we suggest several ways to write
the arithmetic mean. We find the arithmetic mean of the first ten nonzero
natural numbers. Generally $(a_1 + a_2 + \ldots + a_n)/n$. If all the terms are
integers, the result is given as a fraction of two integers *)
N[Plus@@Range@10/10]   (* If at least one entry is given as a decimal
number or the result is required so, the result appears in decimal form *)
GeometricMean@@Table[Range@n,{n,10,10}]   (* The geometric
mean of some nonnegative numbers, i.e., $\sqrt[n]{a_1 a_2 \ldots a_n}$, $a_i \geq 0$ *)
N@%
HarmonicMean@@Table[Range@n,{n,10,10}]   (* The harmonic
mean of some positive numbers, i.e., $n/(1/a_1 + 1/a_2 + \ldots + 1/a_n)$,
$a_i > 0$ *)
N[%]
True
5.5
$2^{4/5}3^{2/5}5^{1/5}7^{1/10}$
4.52873
$\frac{25200}{7381}$
3.41417

Chapter 3
Basic Steps to *Mathematica*

3.1 Problems in Number Theory, Symbolic Manipulation, and Calculus

3.1.1 Problems in Number Theory

3.1.1.1 Pythagorean Numbers

Find all the pairs (n, m) for $1 \leq n \leq m \leq 20$ such that $n^2 + m^2$ is a squared number (*Pythagorean pair* or *Pythagorean numbers*). One approach consists in printing all the satisfactory pairs step by step. If n and m are larger, the resulting list could be rather large.

```
Table[
If[IntegerQ@√(n² + m²)==True,   (* IntegerQ tests whether its argument is
explicitly an integer *)
Print["n = ",n,", m = ",m, ", √(n² + m²)=", √(n² + m²)]],{n,20},{m,n,20}];
```

(*We produce pairs of integers with components from 1 to 20 and test
if each pair is a Pythagorean one. If we find a Pythagorean pair, display
its members and the square root of the sum of squares of members.
Obviously, we take into account the commutativity of the operations *)
$n = 3, m = 4, \sqrt{n^2 + m^2} = 5$
$n = 5, m = 12, \sqrt{n^2 + m^2} = 13$
$n = 6, m = 8, \sqrt{n^2 + m^2} = 10$
$n = 8, m = 15, \sqrt{n^2 + m^2} = 17$
$n = 9, m = 12, \sqrt{n^2 + m^2} = 15$
$n = 12, m = 16, \sqrt{n^2 + m^2} = 20$
$n = 15, m = 20, \sqrt{n^2 + m^2} = 25$

© Springer International Publishing AG 2017
M. Mureşan, *Introduction to Mathematica® with Applications*,
DOI 10.1007/978-3-319-52003-2_3

Another approach to Pythagorean numbers is introduced below [33, p. 83]:

Table[{n,m},{n,6},{m,n,6}]
Flatten[Table[{n,m},{n,6},{m,n,6}],1]   (* We get a list of pairs *)
Select[Flatten[Table[{n,m},{n,20},{m,n,20}],1]   (* We flatten the list
of pairs and select the Pythagorean numbers. We may consider that
m≤n. The built-in function selects and picks out the pairs that satisfy
the logical condition *)
(Sqrt[#[[1]]^2 + #[[2]]^2 ∈Integers)&]   (* The result appears as a list
of Pythagorean pairs *)
{{{1, 1}, {1, 2}, {1, 3}, {1, 4}, {1, 5}, {1, 6}}, {{2, 2}, {2, 3}, {2, 4}, {2, 5},
{2, 6}}, {{3, 3}, {3, 4}, {3, 5}, {3, 6}}, {{4, 4}, {4, 5}, {4, 6}}, {{5, 5}, {5, 6}},
{{6, 6}}}
{{1, 1}, {1, 2}, {1, 3}, {1, 4}, {1, 5}, {1, 6}, {2, 2}, {2, 3}, {2, 4}, {2, 5}, {2, 6},
{3, 3}, {3, 4}, {3, 5}, {3, 6}, {4, 4}, {4, 5}, {4, 6}, {5, 5}, {5, 6}, {6, 6}}
{{3, 4}, {5, 12}, {6, 8}, {8, 15}, {9, 12}, {12, 16}, {15, 20}}

The above result on Pythagorean numbers can be given as a table using the built-in function Grid. We display the triplet $\left(n, m, \sqrt{n^2 + m^2}\right)$, where (n, m) are some Pythagorean numbers.

header={n,m," $\sqrt{n^2 + m^2}$" };   (* This is the header of the grid. Thus, we
have three columns *)
f[{x_,y_}]:={x,y, $\sqrt{x^2 + y^2}$ };   (* The rule by which we add the third
column *)
Grid[Join[{header},f/@Select[Flatten[Table[{n,m},{n,20},{m,n,20}],1],
(Sqrt[#[[1]]^2 + #[[2]]^2] ∈Integers)&]],
Frame→All, Alignment→Right,FrameStyle→Thin]   (* f/@ expr is
equivalent to Map[f, expr] and applies f to each element on the first level
in *expr*. Then the header is added on the first position of the list, and
finally the list is displayed *)

n	m	$\sqrt{n^2 + m^2}$
3	4	5
5	12	13
6	8	10
8	15	17
9	12	15
12	16	20
15	20	25

3.1.1.2   Euler's Sum of Powers Conjecture

Euler's conjecture was proposed in 1769. It states that for all integers n and k greater than 1, if the sum of n kth powers of positive integers is itself a kth power, then n

is greater than or equal to k, i.e., $\sum_{i=1}^{n} a_i^k = b^k$, where $n > 1$ and $a_1, a_2, \ldots, a_n, b$ are positive numbers, then $n \geq k$. He also provided a complete solution to the four cubes problem $3^3 + 4^3 + 5^3 = 6^3$, i.e.,

$3^3 + 4^3 + 5^3 == 6^3$

True

We will study some particular cases.

```
Do[
Do[
Do[
If[IntegerQ@ ∛(i³ + j³ + k³) == True, Print["i = ", i, ", j = ", j, ", k = ", k]],
{k,j,20}],
{j,i,15}],
{i,10}]
```
i = 1, j = 6, k = 8
i = 2, j = 12, k = 16
i = 3, j = 4, k = 5
i = 3, j = 10, k = 18
i = 6, j = 8, k = 10
i = 7, j = 14, k = 17
i = 9, j = 12, k = 15

We can approach this exercise from another point of view.

```
header3={i, j, k, " ∛(i³ + j³ + k³)"};    (* This is the header of the grid *)
f3[{x_,y_,z_}]:={x, y, z, ∛(x³ + y³ + z³)};    (* It transforms a triplet into a
four-component list *)
triple={};   (* The list of triples is empty at the beginning *)

Do[Do[Do[
If[IntegerQ@ ∛(i³ + j³ + k³) == True, AppendTo[triple, {i, j, k}]],
{k,j,30}],{j,i,25}],{i,10}]    (* The commutativity allows us to consider
only the cases 1 ≤ i ≤ j ≤ k *)
Grid[Join[{header3},f3/@triple],Frame→All,Alignment→Right,
FrameStyle→Thin]
```

i	j	k	$\sqrt[3]{i^3 + j^3 + k^3}$
1	6	8	9
2	12	16	18
3	4	5	6
3	10	18	19
3	18	24	27
4	17	22	25
6	8	10	12
7	14	17	20
9	12	15	18
11	15	27	29
12	16	20	24
15	20	25	30
18	19	21	28
18	24	30	36

On the same line, we have $30^4 + 120^4 + 272^4 + 315^4 = 353^4$, i.e.,

$30^4 + 120^4 + 272^4 + 315^4 == 353^4$
True

Table[If[IntegerQ@$\sqrt[4]{i^4 + j^4 + k^4 + m^4}$==True,
Print["i = ",i,", j = ",j,", k = ",k,", m = ",m]],
{m,300,315},{k,250,272},{j,100,120},{i,25,30}];
i = 30, j = 120, k = 272, m = 315

For $k = 5$, the conjecture was disproved by L. J. Lander and T. R. Parkin in 1966 [39] when they found the following counterexample $27^5 + 84^5 + 110^5 + 133^5 = 144^5$, i.e.,

$27^5 + 84^5 + 110^5 + 133^5 == 144^5$
True

For $k = 4$, Noam Elkies in [22] found a counterexample which states that $2682440^4 + 15365639^4 + 18796760^4 = 20615673^4$, i.e.,

$2682440^4 + 15365639^4 + 18796760^4 == 20615673^4$
True

In 1988, Roger Frye in [27] subsequently found the smallest possible counterexample for k = 4: $95800^4 + 217519^4 + 414560^4 = 422481^4$, i.e.,

$95800^4 + 217519^4 + 414560^4 == 422481^4$
True

3.1.1.3 A Conjecture of Fermat

In 1640, Fermat stated that every number of the form $2^{(2^n)} + 1$ is a prime number. Almost 200 years later, an example contradicted Fermat's statement.

Table[PrimeQ[2^{2^n}],{n,10}]   (* PrimeQ offers an answer whether its argument is prime or not; then the corresponding Boolean value is assigned to a list *)

{True,True,True,True,False,False,False,False,False,False}

Table[If [PrimeQ [$2^{2^n} + 1$] == False, Print ["for n = ", n, ", number = $2^{2^n} +1 =$", $2^{2^n} + 1$, "is not a prime"]], {n, 10}];   (* Display the nonprime numbers of the form $2^{2^n} + 1$ for all n∈ {1,2,...,10} *)

FactorInteger[$2^{2^5} + 1$]   (* For a nonprime number, we display its decomposition in powers of primes *)

FactorInteger[$2^{2^6} + 1$]

for n = 5, number = 2^{2^n}+1 = 4294967297 is not a prime

for n = 6, number = 2^{2^n}+1 = 18446744073709551617 is not a prime

for n = 7, number = 2^{2^n}+1 = 340282366920938463463374607431768211457 is not a prime

for n = 8, number = 2^{2^n}+1 = 115792089237316195423570985008687907853 2699846665640564039457584007913129639937 is not a prime

for n = 9, number = 2^{2^n}+1 = 134078079299425970995740249982058461274 793658205923933777235614437217640300735469768018742981669034 2 76900318581864860508537538828119465699464336490006084097 is not a prime

for n = 10, number = 2^{2^n}+1 = 179769313486231590772930519078902473 3 61797697894230657273430081157732675805500963132708477322407536 02112011387987139335765878976881441662249284743063947412437776 78934248654852763022196012460941194530829520850057688381506823 42462881473913110540827237163350510684586298239947245938479716 30483535632962422413721 7 is not a prime

{{641, 1}, {6700417, 1}}

{{274177, 1}, {67280421310721, 1}}

Now we recompose the number from FactorInteger and compare it with the initial number.

power[{x_,y_}]:=Power[x,y];
Times@@power/@FactorInteger[$2^{2^5} + 1$]==$2^{2^5} + 1$
True

3.1.2 Symbolic Manipulations

The binomial coefficient Binomial[n, k] is the number of ways of picking k unordered elements from a set of n elements. By definition Binomial[n,0]=Binomial[n,n]=1.

Binomial[m+1,n+1]==Binomial[m,n+1]+Binomial[m,n]   (* This is the
recurrence of binomials; no hint about its validity *)
FullSimplify[Binomial[m,n]+Binomial[m,1+n]]   (* We can overcome
the previous situation by using the FullSimplify built-in function *)
Binomial[1+m,1+n]==Binomial[m,n]+Binomial[m,1+n]
Binomial[1+m,1+n]

Now we verify for some values of *n* the binomial identity $\sum_{k=0}^{n}$ Binomial[n, k]
$= 2^n$.

sumbinomials[n_]:=1+Plus@@Binomial[n,Range@n]   (* Define the
sum $\sum_{k=0}^{n}$Binomial[n,k] *)
Table[Boole[sumbinomials[n]==2^n],{n,10}]   (* Test if the sum is equal
to the corresponding power of 2. If yes, it inserts "1" in the list,
otherwise inserts "0" *)
With[{m=10},
If[Times@@Table[Boole[sumbinomials[n]==2^n],{n,m}]==1,
Print["True"],Print["False"]]]
{FullSimplify[$\sum_{k=0}^{v}$Binomial[v,k]],FullSimplify[$\sum_{k=0}^{v}$Binomial[v,2k]],
FullSimplify[$\sum_{k=0}^{v}$Binomial[v,2k+1]]   (* Here we calculate some sums
and simplify the results *)
FullSimplify[{$\sum_{k=0}^{v}$Binomial[v,k], $\sum_{k=0}^{v}$Binomial[v,2k],
$\sum_{k=0}^{v}$Binomial[v,2k+1]}]   (* Here there are the previous computations
written in a shorter form *)}
{1, 1, 1, 1, 1, 1, 1, 1, 1, 1}
True
$\{2^v, 2^{-1+v}, 2^{-1+v}\}$
$\{2^v, 2^{-1+v}, 2^{-1+v}\}$

Some binomial identities are exhibited below:

Clear[n,m]
FullSimplify[{$\sum_{m=0}^{k}$Binomial[n+m,m],$\sum_{m=0}^{k}$Binomial[n+m,n],
$\sum_{m=1}^{n}$ m Binomial[n,m],$\sum_{k=0}^{n}$Binomial[n+k,k]Divide[1,2^k],
$\sum_{i=m}^{n}$Binomial[n,i] Binomial[i,m],$\sum_{i=0}^{m}(-1)^i$Binomial[n,i]}]
$\{1, 1, 2^{-1+n}\ n, 2^n, 2^{-m+n}$Binomial[n,m], $(-1)^m$Binomial[-1+n,m]\}$

Now we take a look at the *Bell numbers*. Set $n \in \mathbb{N}$. The *n*th Bell number is the
number of partitions of an *n*-set. Its corresponding *Mathematica* built-in function is
BellB[n].

sumBell[n_]:=1+$\sum_{k=1}^{n-1}$Binomial[n-1,k]*sumBell[k]   (* We define our
Bell   numbers   based   on   the   corresponding   recurrence   *)
Table[sumBell[n],{n,1,10}] (* The first ten Bell numbers are displayed *)
Table[Boole[sumBell[n]==BellB[n]],{n,10}]   (* We test if our Bell
numbers coincide with the built-in Bell numbers only for the first ten terms *)

```
With[{m=10},
If[Times@@Table[Boole[sumBell[n]==BellB[n]],{n,m}]==1,
Print["True"],
Print["False"]]]
{1, 2, 5, 15, 52, 203, 877, 4140, 21147, 115975}
{1, 1, 1, 1, 1, 1, 1, 1, 1, 1}
True
```

Another example of symbolic manipulation requires checking whether $\tan(3\pi/11) + 4\sin(2\pi/11) = \sqrt{11}$.

```
Simplify[Tan[3π/11]+4Sin[2π/11]]   (* The built-in function Simplify is of
no help here *)
FullSimplify[Tan[3π/11]+4Sin[2π/11]]   (* The built-in function
FullSimplify is of real help *)
```
$\mathrm{Cot}\left[\frac{5\pi}{22}\right] + 4\mathrm{Sin}\left[\frac{2\pi}{11}\right]$
$\sqrt{11}$

By the next example, we show that $\sqrt{\sqrt[3]{64}\,(2^2 + (1/2)^2) - 1} = 4$.

```
Simplify[ √(∛64(2² + (1/2)²) − 1) ]   (* The result immediately follows *)
4
```

We show that $e^{I\pi} = -1$.

```
Exp[Iπ]   (* The result follows at once *)
−1
```

3.1.3 Texts

We introduce an example of a text. Suppose the word "Mathematica" is lying on the graph of the sine function.

```
Show[Graphics[MapThread[Text[StyleForm[#1,FontFamily→"Courier",
FontWeight→"Bold",FontSize→14,FontColor→Hue@#2],
{#2,Sin@#2}]&,{Characters["Mathematica"],Table[2πi/11,{i,11}]}]]]
```

See Fig. 3.1.

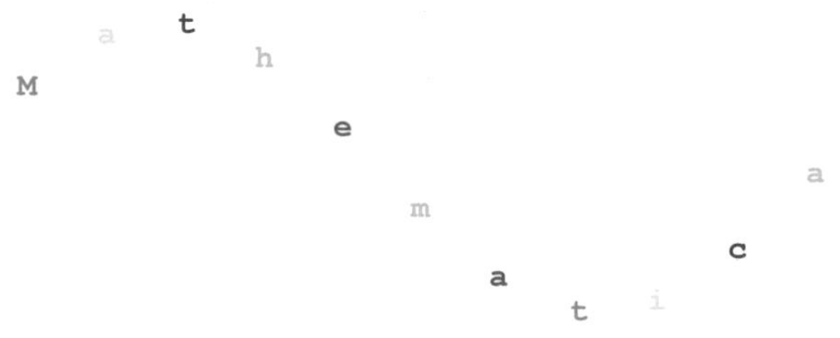

Fig. 3.1 Mathematica on the sine function

An interesting example with texts similar to the one mentioned earlier may be found in [65, pp. 6–7].

3.2   Riemann ζ Function

Riemann's zeta function $\zeta(p)$ is defined by the relation $\zeta(p) = \sum_{k=1}^{\infty} k^{-p}$ (for $p > 1$).

Riemann ζ function at point 2 is

Zeta@2

$\frac{\pi^2}{6}$

We get the same result as a limit

$\text{Limit}\left[\sum_{k=1}^{n} \frac{1}{k^2}, n \to \text{Infinity}\right]$

$\frac{\pi^2}{6}$

Here there are the values of the Riemann ζ function corresponding to some values of the argument p.

Table[Zeta@p,{p,9}]

$\text{Table}\left[\text{Limit}\left[\sum_{k=1}^{n} \frac{1}{k^p}, n \to \infty\right], \{p,9\}\right]$ (* We look for the values of Riemann ζ function as limits *)

$\text{Table}\left[\text{N}\left[\text{Limit}\left[\sum_{k=1}^{n} \frac{1}{k^p}, n \to \infty\right]\right], \{p, 3, 9, 2\}\right]$

$\left\{\text{ComplexInfinity}, \frac{\pi^2}{6}, \text{Zeta}[3], \frac{\pi^4}{90}, \text{Zeta}[5], \frac{\pi^6}{945}, \text{Zeta}[7], \frac{\pi^8}{9450}, \text{Zeta}[9]\right\}$

$\left\{\infty, \frac{\pi^2}{6}, \text{Zeta}[3], \frac{\pi^4}{90}, \text{Zeta}[5], \frac{\pi^6}{945}, \text{Zeta}[7], \frac{\pi^8}{9450}, \text{Zeta}[9]\right\}$

$\{1.20206, 1.03693, 1.00835, 1.00201\}$

3.3   Some Numerical Sequences

3.3.1   The First Sequence

Let us analyze the sequence $a_n = \sqrt[n]{n}, n \in \mathbb{N}, n \geq 2$.

Table[N@$\sqrt[n]{n}$, {n, 20}]   (* The resulting list suggests that the sequence is eventually decreasing *)

$\{1., 1.41421, 1.44225, 1.41421, 1.37973, 1.34801, 1.32047, 1.29684, 1.27652,$
$1.25893, 1.24358, 1.23008, 1.21811, 1.20744, 1.19786, 1.18921, 1.18135,$
$1.17419, 1.16762, 1.16159\}$

We check it by the next code.

Limit $\left[\{\text{N}@\sqrt[n]{n}, \sqrt[n]{n}\}, n \to \infty\right]$ (* The limit is 1 *)

$\{1, 1\}$

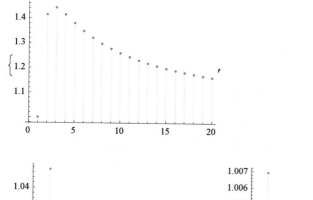

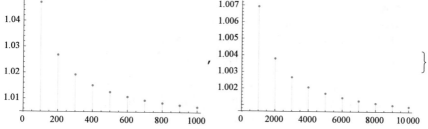

Fig. 3.2 The sequence $\sqrt[n]{n}$

We want to visualize its convergence toward 1.

$$\text{ListPlot}\left[\text{Table}\left[\{n, N@\sqrt[n]{n}\}, \#\right], \text{Filling} \to \text{Axis}, \text{ImageSize} \to 200\right] \&/@$$
$$\{\{n, 1, 20\}, \{n, 100, 1000, 100\}, \{n, 10^3, 10^4, 10^3\}\}$$

The graphs suggest the behavior of this sequence; see Fig. 3.2.

3.3.2 The Second Sequence

We consider the sequence defined

$$a_n = \sqrt[n+1]{(n+1)!} - \sqrt[n]{n!}, \quad n \in \mathbb{N}^*.$$

A classical solution to it may be found at [53, pp. 106–107]. With *Mathematica* the solution is clear immediately, considering the next code.

$$\text{Assuming}\left[n \in \text{Integers}\&\&n > 0, \text{Limit}\left[\sqrt[n+1]{(n+1)!} - \sqrt[n]{n!}, \ n \to \infty\right]\right]$$
$$\frac{1}{e}$$

3.3.3 The Third Sequence

We now look at the convergence of a sequence defined by a recurrence [53, p. 141]. The sequence is defined by

$$x_{n+1} = 1 - x_n^2, \quad x_1 = a \in\,]0, 1[, \quad n \in \mathbb{N}^*.$$

We take a look at the behavior of the terms of the sequence considering three different initial values $.25, .9,$ and $\left(-1 + \sqrt{5}\right)\big/ 2$. The following code is helpful:

```
With[{m=7},   (* Number of terms *)
x[1]=.25;x[n_]:=1-x[n-1]^2;   (* The first term and the first recurrence *)
y[1]=.9;y[n_]:=1-y[n-1]^2;   (* The first term and the second recurrence *)
z[1]=(-1+Sqrt[5])/2;z[n_]:=1-z[n-1]^2;   (* The first term and the
third recurrence *)
header={"index","sequence","sequence","sequence"};
{Grid[Join[{header},Table[{k,x[k],y[k],Simplify[z[k]]},{k,m}]],
Frame→All,Alignment→{{Right,Left,Left,Left}}],
data1=Table[{n,x[n]},{n,m}];data2=Table[{n,y[n]},{n,m}];
data3=Table[{n,z[n]},{n,m}];
ListLinePlot[{Tooltip[data1],Tooltip[data2],Tooltip[data3]},
PlotStyle→{Blue,Red,Darker[Green]},
Mesh→Full,MeshStyle→{Black,PointSize[Medium]},
ImageSize→{280,200},Ticks→{Table[μ,{μ,m}],{{.25,".25"},
{(-1+Sqrt[5])/2, " -1+Sqrt[5]/2 "},{.9,".9"},1}}]}]
```

The results are exhibited in Fig. 3.3.

Denote $\varepsilon = (-1 + \sqrt{5})/2 \in\,]0, 1[$. We have the following cases:

(i) $0 < x_n < 1 \Longrightarrow 0 < x_{n+1} < 1$. Thus, the sequence is lying in the open interval $]0, 1[$.

n	x[n]	y[n]	z[n]
1	0.25	0.9	$\frac{1}{2}\left(-1+\sqrt{5}\right)$
2	0.9375	0.19	$\frac{1}{2}\left(-1+\sqrt{5}\right)$
3	0.121094	0.9639	$\frac{1}{2}\left(-1+\sqrt{5}\right)$
4	0.985336	0.0708968	$\frac{1}{2}\left(-1+\sqrt{5}\right)$
5	0.0291124	0.994974	$\frac{1}{2}\left(-1+\sqrt{5}\right)$
6	0.999152	0.0100274	$\frac{1}{2}\left(-1+\sqrt{5}\right)$
7	0.00169434	0.999899	$\frac{1}{2}\left(-1+\sqrt{5}\right)$

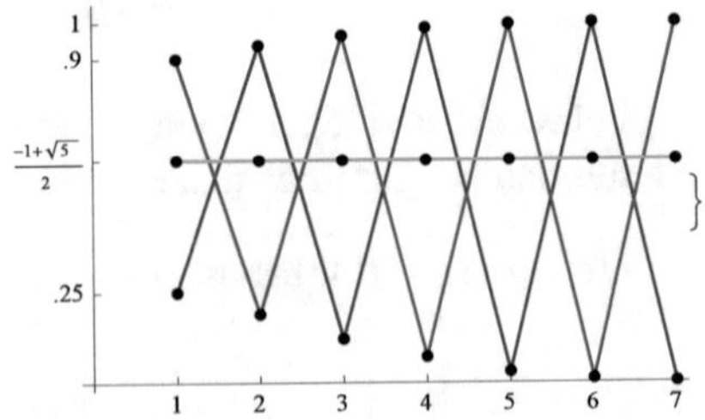

Fig. 3.3 The third sequence

(ii) Because of $\text{sgn}(x_{n+1} - x_n) = -\text{sgn}(x_n - \varepsilon)$, it implies that if $x_n = \varepsilon$, then $x_{n+1} = \varepsilon$, and thus, the sequence is constant.

(iii) Because of $\text{sgn}(x_{n+1} - \varepsilon) = -\text{sgn}(x_n - \varepsilon)$, it implies that the sequence oscillates if $x_1 \neq \varepsilon$. We also have $|x_{n+2} - x_{n+1}| > |x_{n+1} - x_n|$, which implies that $|x_{n+2} - x_{n+1}| > |x_2 - x_1|$, so the convergence is excluded.

Thus, the sequence converges in a single case and diverges in all the other cases.

3.4 Variables

Just like in mathematics, we need and use variables of different sorts and for different purposes.

The next command sets the value of the variable x to 7.

```
x=7
7
```

If we do not want to display this number, we write

```
x = 7;   (* A semicolon ";" stops printing the value of the variable *)
N@Power[x, (7)^{-1}]   (* Instead, we want to know the numerical value
of ⁷√7 *)
1.32047
```

Now we can use this variable in computations such as

```
x+E^x+Log@x+Sin@x+Tan@x+Floor[x+.25]+Ceiling[x-.25]
21+e^7+Log[7]+Sin[7]+Tan[7]
```

In certain cases, if we want a numerical result, then we write

```
N[%,12]   (* % means that the previous result is taken into account
and we ask for a result of 12 *)
1121.10750316
```

If the input variable is a non-integer real number, we immediately get the desired result with 5 decimals without using the built-in function N.

```
x=7.00001
x+E^x+Log@x+Sin@x+Tan@x+Floor[x+.25]+Ceiling[x-.25]
7.00001
1121.12
```

As we have already seen, if we add a semicolon (;) at the end of an insertion line, the result is not returned, but it is calculated and kept.

```
x=2.00001;
x+E^x+Log@x+Sin@x+Tan@x+Floor[x+.25]+Ceiling[x-.25]
N[%,8]
12.8066
```

3.5   Lists

A *list* is a sequence of objects. It is not a set (in a mathematical sense) because the
rank of an element in a list counts, which is not the case of a set. A proper synonym
of list (in the sense of *Mathematica*) is sequence (in mathematical sense).

3.5.1   *Operations with Lists*

Here are two lists of numbers having the same number of elements. We add and
subtract them, rise to power, and perform other operations. The operations are
performed componentwise.

```
aa={1,3,4,2,-3};   (* The first list *)
bb={2,-3,8,-7,1};   (* The second list *)
{aa+bb,   (* In order to save space, the results of the operations on the
two lists appear as members of a list *)
bb-aa}
{aa²+1,   (* The operations are performed componentwise *)
Exp[%]//N,
aa/(bb-1.5),   (* Division is operated componentwise *)
xx{1,4,-11},
{1,2,5.1}{1,4,-11}   (* The two lists must have the same number of
elements *)}
```

$\{\{3, 0, 12, -5, -2\}, \{1, -6, 4, -9, 4\}\}$
$\{\{2, 10, 17, 5, 10\}, \{\{20.0855, 1., 162755., 0.00673795, 0.135335\},$
$\{2.71828, 0.00247875, 54.5982, 0.00012341, 54.5982\}\},$
$\{2., -0.666667, 0.615385, -0.235294, 6.\}, \{xx, xx^4, \frac{1}{xx^{11}}\},$
$\{1, 16, 1.647130704063058` * 10^{-8}\}\}$

We often need to know the values of certain elements in a list. There are two
ways to extract some elements in a list.

```
a={1,4,5,2,-7}
{a[[3]],   (* We extract the third element indicating the rank of the chosen
element *)
Part[a,4],   (* We extract the fourth element *)
a[[{5,3}]],   (* We extract the fifth and third element, in this order.
The result is a list of two elements *)
Part[a,{5,1}],   (* We extract the fifth and first element, in this order.
Again the result is a list *)
a[[-2]]   (* It counts from the end, i.e., the second element from the
end *)}
```

$\{1, 4, 5, 2, -7\}$
$\{5, 2, \{-7, 5\}, \{-7, 1\}, 2\}$

A list can be rearranged as the following example shows:

```
Clear[a]
a={1,4,9,16};
a[[{4,3,2,1}]]
{16,9,4,1}
```

The previous rearrangement can be performed by the next command.

```
a[[Table[Length@a-k,{k,0,Length@a-1}]]]
{16,9,4,1}
```

We now introduce a list of lists.

```
Clear[a]
a={-1,1,{2,3},{α,β,γ},{μ,{ν,π,bird}}}
{-1,1,{2,3},{α,β,γ},{μ,{ν,π,bird}}}
```

Then

```
{a[[3]],    (* We pick the third element which is a list *)
a[[3,1]],   (* From the third element, we pick the first *)
a[[4]],a[[4,{1,2}]],   (* From the fourth element, we pick the first and
the second *)
a[[5]],a[[5,2]],a[[5,2]][[2]]}   (* From the fifth element we pick the
second and from it again the second *)
{{2,3},2,{α,β,γ},{α,β},{μ,{ν,π,bird}},{ν,π,bird},π}
```

```
Clear[a]
a={1,4,9,13}
Length@a   (* It returns the length of the list *)
{First@a,   (* It extracts the first element *)
Last@a,   (* It extracts the last element *)
Drop[a,3],   (* It gives the list with its first three elements dropped *)
Drop[a,{2,4}],   (* It gives the list with the second until the fourth
elements dropped *)
Rest@a,   (* The first element is removed *)
Most@a,   (* The last element is removed *)
Take[a,{2,4}],   (* It gives the specified elements in a list *)
Most@Rest@a==Rest@Most@a}   (* We ask if the left-hand side
coincides with the right-hand side *)
{1,4,9,13}
4
{1,13,{13},{1},{4,9,13},{1,4,9},{4,9,13},True}
```

```
{Riffle[a,picture],   (* It inserts the variable picture between successive
elements of the list a *)
Riffle[a,{x,y}],   (* It inserts x and y cyclically *)
Riffle[{aa,bb,cc},{x,y,z}]}
{{1,picture,4,picture,9,picture,13},{1,x,4,y,9,x,13},{aa,x,bb,y,cc,z}}
```

Clear[b,c,d]
{First$[c^2 + c + b^2 + b + d^2]$,   (* The elements are sorted ascending;
then the first element is selected *)
Last$[c^2 + b^2 + d^2 + d]$,   (* The elements are sorted ascending; then
the last element is selected *)
Drop$[c^2 + b^2 + d^2, 2]$,   (* The elements are sorted ascending; then
the first two elements are dropped *)
Rest$[c^2 + b^2 + d^2]$,   (* The elements are sorted ascending; then the
first element is removed *)
Most$[b^2 + d^2 + c]$,   (* The elements are sorted ascending; then the last
element is removed *)}
$\{b, d^2, d^2, c^2 + d^2, b^2 + c\}$

a={{1,4},9,{13,{14,15}}}
Flatten@a   (* It flattens out the nested list *)
Flatten[{a[[2]],a[[3]],a[[1]]}]
{{1,4},9,{13,{14,15}}}
{1,4,9,13,14,15}
{9,13,14,15,1,4}

Clear[b,c]
FlattenAt[{a,{b,c},{d,e},{f}},{{2},{4}}]   (* FlattenAt flattens the
positions 2 and 4 *)
{{1,4,9,13},b,c,{d,e},f}

A list can be rearranged as the following examples show. We use the built-in
function Sort and change the places of certain elements according to some criteria.

Clear[a]
a={1,9,4,4,16};
Sort@a   (* Implicitly it sorts ascending *)
Sort[a,Greater]   (* This sorts in descending order *)
Sort[a,#1>#2&]   (* We ask for a descending order *)
Sort[{{a,2},{c,1},{d,3}},#1[[2]] < #2[[2]]&]   (* It sorts ascending with
respect to the second component *)
Sort[{{a,2},{c,1},{d,3}},#1[[1]] < #2[[1]]&]   (* It sorts ascending with
respect to the first component *)
{1,4,4,9,16}
{16,9,4,4,1}
{16,9,4,4,1}
{{c,1},{{1,9,4,4,16},2},{d,3}}
{{{1,9,4,4,16},2},{c,1},{d,3}}

A deeper discussion on the sorting problem will be given in the next chapter.
Some elements of a list can be reset. Let us reset the third element of the list.

```
u={1,3,9};
u[[3]]=♡;
u
{1,3,♡}
```

Manipulation of lists.

```
Clear[a,b,c,d,e,x,y,t,u]
Join[{a,c,b},{x,y},{t,a}]   (* Join concatenates lists. The result is an
unsorted list *)
{a,c,b,x,y,t,a}
```

```
Union[{a,c,b},{b,c,d},{c,d,e}]   (* Union combines lists, keeping only
distinct elements. It supplies a list sorted ascending *)
{a,b,c,d,e}
```

```
Intersection[{1,7,2,3},{3,1,2},{2,1,3,3,π}]   (* Intersection gives a
sorted list of the elements common to all the lists. It supplies a
list sorted ascending *)
{1,2,3}
```

```
Complement[{11,7,2,3},{3,1,2},{2,1,π}]   (* Complement gives the list
of elements in the first set not belonging to any other set. The list is
sorted ascending *)
{7,11}
```

Suppose we want to see if all elements of a sequence are the same.

```
a={1,1,1};
Length@Union@a==1
a={1,1,2};
Length@Union@a==1
True
False
```

Another approach is the following: http://community.wolfram.com/groups/-/m/
t/767736?_19.

```
a={1,1,1};
Equal@@a
a={1,1,2};
Equal@@a
True
False
```

A list of the form {{1, 2}, {3, 4}} can be considered a 2 × 2 matrix.

```
ma={{1,2},{3,4}}//MatrixForm
```
$$\begin{pmatrix} 1 & 2 \\ 3 & 4 \end{pmatrix}$$

From a simple matrix, we can compose more complicated matrices.

m={{1,2},{3,4}}
ArrayFlatten[{{0,0,m},{m,m,0}}]//MatrixForm   (* ArrayFlatten
creates a single flattened matrix from a matrix of matrices *)
{{1,2},{3,4}}

$$\begin{pmatrix} 0 & 0 & 0 & 0 & 1 & 2 \\ 0 & 0 & 0 & 0 & 3 & 4 \\ 1 & 2 & 1 & 2 & 0 & 0 \\ 3 & 4 & 3 & 4 & 0 & 0 \end{pmatrix}$$

m={{1}}
ArrayFlatten[{{m,0,0,0},{0,m,0,0},{0,0,m,0},{0,0,0,m}}]//MatrixForm
{{1}}

$$\begin{pmatrix} 1 & 0 & 0 & 0 \\ 0 & 1 & 0 & 0 \\ 0 & 0 & 1 & 0 \\ 0 & 0 & 0 & 1 \end{pmatrix}$$

{Table[a[i,j]=0,{i,4},{j,4}]//MatrixForm,   (* The null matrix *)
Table[a[i,i]=1,{i,4}]//MatrixForm,   (* The column matrix *)
Table[a[i,j],{i,4},{j,4}]//MatrixForm,   (* The identity or unit matrix *)
DiagonalMatrix[{a,b,c,d}]//MatrixForm   (* A matrix with prescribed
diagonal elements, all the other elements are null *)}

$$\left\{ \begin{pmatrix} 0 & 0 & 0 & 0 \\ 0 & 0 & 0 & 0 \\ 0 & 0 & 0 & 0 \\ 0 & 0 & 0 & 0 \end{pmatrix}, \begin{pmatrix} 1 \\ 1 \\ 1 \\ 1 \end{pmatrix}, \begin{pmatrix} 1 & 0 & 0 & 0 \\ 0 & 1 & 0 & 0 \\ 0 & 0 & 1 & 0 \\ 0 & 0 & 0 & 1 \end{pmatrix}, \begin{pmatrix} a & 0 & 0 & 0 \\ 0 & b & 0 & 0 \\ 0 & 0 & c & 0 \\ 0 & 0 & 0 & d \end{pmatrix} \right\}$$

3.5.2   *Operations with Matrices*

In this subsection, we will take a look at some elementary operations on matrices
[1, 11]. For that, we introduce three sample matrices. As mentioned above, a matrix
is a list of lists, the lists having the same number of elements.

ma={{0,0,-2,1},{3,-1,7,2},{-6,0,5,-1},{-6,0,1,-2}};
mb={{-7,-6,-3,-7},{2,-3,0,4},{3,4,1,2},{5,6,3,6}};
mc={{-7,-6,-3,-7,-2},{2,-3,0,4,2},{3,4,1,2,0},{5,6,3,6,-2}};
{MatrixForm@ma,   (* Here we have the matrix ma under the
classical form, that is, a rectangular array of numbers, symbols, or
expressions *)
MatrixForm@mb,
MatrixForm@mc}

$$\left\{ \begin{pmatrix} 0 & 0 & -2 & 1 \\ 3 & -1 & 7 & 2 \\ -6 & 0 & 5 & -1 \\ -6 & 0 & 1 & -2 \end{pmatrix}, \begin{pmatrix} -7 & -6 & -3 & -7 \\ 2 & -3 & 0 & 4 \\ 3 & 4 & 1 & 2 \\ 5 & 6 & 3 & 6 \end{pmatrix}, \begin{pmatrix} -7 & -6 & -3 & -7 & -2 \\ 2 & -3 & 0 & 4 & 2 \\ 3 & 4 & 1 & 2 & 0 \\ 5 & 6 & 3 & 6 & -2 \end{pmatrix} \right\}$$

We calculate $3ma - 2mb$, mb^T, $ma \times mb$, $|ma|$, $|mb|$, $|ma \times mb|$, and ma^{-1}.

```
matrix1=3ma-2mb//MatrixForm;
matrix2=Transpose@mb//MatrixForm;
matrix3=ma.mb;   (* Product of matrices *)
deta=Det@ma;   (* Determinant of a matrix *)
detb=Det@mb;
detab=Det@matrix3;
matrix7=Inverse@ma//MatrixForm;   (* The inverse of a matrix *)
{matrix1,matrix2}
{matrix3//MatrixForm,deta,detb,detab,detab==deta × detb}
matrix7
```

$$\left\{ \begin{pmatrix} 14 & 12 & 0 & 17 \\ 5 & 3 & 21 & -2 \\ -24 & -8 & 13 & -7 \\ -28 & -12 & -3 & -18 \end{pmatrix}, \begin{pmatrix} -7 & 2 & 3 & 5 \\ -6 & -3 & 4 & 6 \\ -3 & 0 & 1 & 3 \\ -7 & 4 & 2 & 6 \end{pmatrix} \right\}$$

$$\left\{ \begin{pmatrix} -1 & -2 & 1 & 2 \\ 8 & 25 & 4 & 1 \\ 52 & 50 & 20 & 46 \\ 35 & 28 & 13 & 32 \end{pmatrix}, -36, -24, 864, \text{True} \right\}$$

$$\begin{pmatrix} -\frac{1}{4} & 0 & -\frac{1}{12} & -\frac{1}{12} \\ -\frac{7}{12} & -1 & \frac{19}{12} & -\frac{25}{12} \\ -\frac{1}{6} & 0 & \frac{1}{6} & -\frac{1}{6} \\ \frac{2}{3} & 0 & \frac{1}{3} & -\frac{1}{3} \end{pmatrix}$$

We want to find the eigenvalues of a matrix ma.

```
ma={{0,0,-2,1},{3,-1,7,2},{-6,0,5,-1},{-6,0,1,-2}};
Eigenvalues@ma   (* Eigenvalues computed by the built-in function
Eigenvalues *)
soleingen=Solve[Det[ma-λIdentityMatrix[4]]==0]   (* Eigenvalues
computed based on definition *)
N@soleingen
```

$\{\text{Root}\left[-36 - 15\#1 - 3\#1^2 + \#1^3 \&, 1\right],$
$\text{Root}\left[-36 - 15\#1 - 3\#1^2 + \#1^3 \&, 3\right],$
$\text{Root}\left[-36 - 15\#1 - 3\#1^2 + \#1^3 \&, 2\right], -1\}$

$\left\{ \{\lambda \to -1\}, \left\{ \lambda \to 1 + \frac{1}{3}\left(\frac{1431}{2} - \frac{27\sqrt{1945}}{2}\right)^{1/3} + \left(\frac{1}{2}\left(53 + \sqrt{1945}\right)\right)^{1/3} \right\}, \right.$

$$\left\{\lambda \to 1 - \tfrac{1}{6}\left(1 + i\sqrt{3}\right)\left(\tfrac{1431}{2} - \tfrac{27\sqrt{1945}}{2}\right)^{1/3} - \right.$$
$$\left. \tfrac{1}{2}\left(1 - i\sqrt{3}\right)\left(\tfrac{1}{2}\left(53 + \sqrt{1945}\right)\right)^{1/3}\right\},$$

$$\left\{\lambda \to 1 - \tfrac{1}{6}\left(1 - i\sqrt{3}\right)\left(\tfrac{1431}{2} - \tfrac{27\sqrt{1945}}{2}\right)^{1/3} - \right.$$
$$\left. \tfrac{1}{2}\left(1 + i\sqrt{3}\right)\left(\tfrac{1}{2}\left(53 + \sqrt{1945}\right)\right)^{1/3}\right\}\right\}$$

$\{\{\lambda \to -1.\}, \{\lambda \to 6.29279\}, \{\lambda \to -1.64639 + 1.735i\},$
$\{\lambda \to -1.64639 - 1.735i\}\}$

We note that the matrix ma has two real and two complex conjugate eigenvalues. The real eigenvalues are

```
soleingen[[{1,2}]]
Part[soleingen,{1,2}]    (* Equivalent to the previous command *)
%//N
{soleingen[[1]],soleingen[[1,1]],soleingen[[1,1,1]],soleingen[[1,1,2]]}
(* Steps to pick the first real eigenvalue *)
```

$$\left\{\{\lambda \to -1\}, \left\{\lambda \to 1 + \tfrac{1}{3}\left(\tfrac{1431}{2} - \tfrac{27\sqrt{1945}}{2}\right)^{1/3} + \left(\tfrac{1}{2}\left(53 + \sqrt{1945}\right)\right)^{1/3}\right\}\right\}$$

$$\left\{\{\lambda \to -1\}, \left\{\lambda \to 1 + \tfrac{1}{3}\left(\tfrac{1431}{2} - \tfrac{27\sqrt{1945}}{2}\right)^{1/3} + \left(\tfrac{1}{2}\left(53 + \sqrt{1945}\right)\right)^{1/3}\right\}\right\}$$

$\{\{\lambda \to -1.\}, \{\lambda \to 6.29279\}\}$
$\{\{\lambda \to -1\}, \lambda \to -1, \lambda, -1\}$

We can perform the previous computations directly, using some built-in functions.

```
ma={{0,0,-2,1},{3,-1,7,2},{-6,0,5,-1},{-6,0,1,-2}};
CharacteristicPolynomial[ma,λ]
{eigv=Eigenvalues[N@ma],eigv[[1]]}
eigvec=Eigenvectors@N@ma
MatrixForm@%[[3]]    (* The third eigenvector *)
```

$-36 - 51\lambda - 18\lambda^2 - 2\lambda^3 + \lambda^4$
$\{\{6.29279 + 0.i, -1.64639 + 1.735i, -1.64639 - 1.735i, -1. + 0.i\},$
$6.29279 + 0.i\}$
$\{\{0.187245 + 0.i, -0.654212 + 0.i, -0.699033 + 0.i, -0.21977 + 0.i\},$
$\{0.0733515 + 0.107752i, 0.9019 + 0.i, -0.0170554 + 0.121983i,$
$-0.341825 + 0.193829i\}, \{0.0733515 - 0.107752i, 0.9019 + 0.i,$
$-0.0170554 - 0.121983i, -0.341825 - 0.193829i\},$
$\{0. + 0.i, 1. + 0.i, 0. + 0.i, 0. + 0.i\}\}$

$$\begin{pmatrix} -0.0733515 & -0.107752i \\ -0.9019 & +0.i \\ 0.0170554 & -0.121983i \\ 0.341825 & 0.193829i \end{pmatrix}$$

3.5.3 Inner and Outer Commands

Given $\{x_1, x_2, \ldots, x_n\}$ and $\{y_1, y_2, \ldots, y_n\}$, how can one produce the lists

$$\{x_1, y_1, x_2, y_2, x_3, y_3, \ldots, x_{n,} y_n\},$$

$$\{\{x_1, y_1\}, \{x_1, y_2\}, \ldots, \{x_1, y_n\}, \{x_2, y_1\}, \{x_2, y_2\}, \ldots, \{x_2, y_n\}, \ldots,$$

$$\{x_n, y_1\}, \{x_n, y_2\}, \ldots, \{x_n, y_n\}\}?$$

The first list can be obtained by the built-in function Riffle or by using the built-in function Inner[f,{a,b},{x,y},g].

Some examples introduced below show the outputs for different uses of the built-in function Inner.

Clear[a,b,c,d,u,v,w,x,y,z]
Inner[f,{a,b,c},{x,y,z},g]   (* This is a quite good definition of the built-in function Inner *)
g[f[a,x],f[b,y],f[c,z]]

{Inner[f,{{a,b},{c,d}},{{u,v},{w,x}},g],
Inner[List,{{a,b},{c,d}},{{u,v},{w,x}},List]}
{{{g[f[a,u],f[b,w]],g[f[a,v],f[b,x]]},{g[f[c,u],f[d,w]],g[f[c,v],f[d,x]]}},
{{{{a,u},{b,w}},{{a,v},{b,x}}},{{{c,u},{d,w}},{{c,v},{d,x}}}}}

{Inner[List,{a,b,c},{x,y,z},List],
Flatten[Inner[List,{a,b,c},{x,y,z},List]],
Inner[Sequence,{a,b,c},{x,y,z},List]   (* It coincides with the previous one *)}
{{{a,x},{b,y},{c,z}},{a,x,b,y,c,z},{a,x,b,y,c,z}}

Inner[Sequence,{a,b,c},{x,y,z},List]==Riffle[{a,b,c},{x,y,z}]   (* It works *)
{Inner[Times,{a,b},{x,y},Plus],   (* Some concrete applications of the built-in function Inner follow *)
Inner[Power,{a,b,c},{x,y,z},Times]}   -
True
{a x+b y, a^x b^y c^z}

mm={{1,2,3},{4,5,6}};   (* The addition of two lists can be performed by the built-in function Inner *)
Inner[Plus,mm[[1]],mm[[2]],List]
{5,7,9}

We study the case of input sequences of arbitrary length (not necessarily equal), i.e., the case Inner[List,list1,list2,List].

```
Clear[a,b]
myInner[a_List,b_List]:=(
n=Length@a;m=Length@b;
list={};min =Min[n,m];max =Max[n,m];
If[n+m>0,
Do[AppendTo[list,a[[k]]];
AppendTo[list,b[[k]]],{k,min}];
If[n<max,Do[AppendTo[list,b[[k]]],{k,min +1,m}],
Do[AppendTo[list,a[[k]]],{k,min +1,n}],
If[m==0,list=a,list=b]]];
Print[list]
)
```

We check our code.

```
x={};y={1,2};   (* The first list is empty *)
myInner[x,y]
x={1};y={};   (* The second list is empty *)
myInner[x,y]
x={3,4};y={1,2};   (* Both lists are nonempty and of equal lengths *)
myInner[x,y]
x={3,4,5};y={1,2};   (* Both lists are nonempty and of different
lengths *)
myInner[x,y]
x={1};y={1,2,7,5};
myInner[x,y]
x={};y={};   (* Both lists are empty *)
myInner[x,y] {1, 2}
{1}
{3, 1, 4, 2}
{3, 1, 4, 2, 5}
{1, 1, 2, 7, 5}
{}
```

Now we focus on the second command, i.e., $Outer[f,list_1,list_2,\dots]$.

```
Clear[a,b,c,d,u,v,w,x,y,z,f]
Outer[f,{a,b},{x,y,z},{α, β}]   (* It is a simple definition of this
command *)
Outer[List,{a,b},{x,y,z},{α, β}]
Flatten[Outer[List,{a,b},{x,y,z},{α, β}],1]   (* It treats only sublists at
level 1 in the lists as separate elements *)
Flatten[Outer[List,{a,b},{x,y,z},{α, β}],2]
Flatten[Outer[List,{a,b},{x,y,z},{α, β}],3]
Flatten[Outer[List,{a,b},{x,y,z},{α, β}],3]==
Flatten[Outer[List,{a,b},{x,y,z},{α, β}]]
```

{{{f[a,x,α],f[a,x,β]},{f[a,y,α],f[a,y,β]},{f[a,z,α],f[a,z,β]}},
{{f[b,x,α],f[b,x,β]},{f[b,y,α],f[b,y,β]},{f[b,z,α],f[b,z,β]}}}
{{{{a,x,α},{a,x,β}},{{a,y,α},{a,y,β}},{{a,z,α},{a,z,β}}},
{{{b,x,α},{b,x,β}},{{b,y,α},{b,y,β}},{{b,z,α},{b,z,β}}}}
{{{a,x,α},{a,x,β}},{{a,y,α},{a,y,β}},{{a,z,α},{a,z,β}},
{{b,x,α},{b,x,β}},{{b,y,α},{b,y,β}},{{b,z,α},{b,z,β}}}
{{a,x,α},{a,x,β},{a,y,α},{a,y,β},{a,z,α},{a,z,β},{b,x,α},
{b,x,β},{b,y,α},{b,y,β},{b,z,α},{b,z,β}}
{a,x,α,a,x,β,a,y,α,a,y,β,a,z,α,a,z,β,b,x,α,b,x,β,b,y,α,b,y,β,b,z,α,b,z,β}
True

Outer[List,{a,b,c,d},{u,v,w},List]   (* If the lists are of different lengths,
the system returns a message *)
Outer::ipnfm : Positive machine-sized integer or infinity expected at position 4 in
Outer[List,{a,b,c,d},{u,v,w},List] ≫
Outer[List,{a,b,c,d},{u,v,w},List]

Note that we can overcome this situation in the following way. Given the lists
$\{x_1, x_2, \ldots, x_n\}$ and $\{y_1, y_2, \ldots, y_m\}$ we want to obtain the next list

$$\{\{x_1, y_1\}, \{x_1, y_2\}, \ldots, \{x_1, y_m\}, \{x_2, y_1\}, \{x_2, y_2\}, \ldots, \{x_2, y_m\}, \ldots,$$

$$\{x_n, y_1\}, \{x_n, y_2\}, \ldots, \{x_n, y_m\}\}$$

Here is the code:

```
Clear[a,b,n,m]
myOuter[a_List,b_List]:=(
n=Length@a;m=Length@b;
If[n×m>0,
list={};k=1;
While[k≤n,
i=1;
While[i≤ m,AppendTo[list,{a[[k]],b[[i]]}];i++];k++];
Print@list;
];
)

x={1,2,6,8};y={4,5,7};
myOuter[x,y]
{{1,4},{1,5},{1,7},{2,4},{2,5},{2,7},{6,4},{6,5},{6,7},{8,4},{8,5},{8,7}}
```

We note that

```
Clear[a,b,c,x,y,z]
Flatten[Outer[List,{a,b,c},{x,y,z}],1]
{{a,x},{a,y},{a,z},{b,x},{b,y},{b,z},{c,x},{c,y},{c,z}}
```

coincides with

```
CartesianProduct[{a,b,c},{x,y,z}]
{{a,x},{a,y},{a,z},{b,x},{b,y},{b,z},{c,x},{c,y},{c,z}}

CartesianProduct[{a,b,c},{x,y,z,t}]   (* Clearly the lists are of
arbitrary lengths *)
{{a,x},{a,y},{a,z},{a,t},{b,x},{b,y},{b,z},{b,t},{c,x},{c,y},{c,z},{c,t}}
```

3.5.4 Again on the Third Sequence

In order to provide a shorter code, we reconsider the third sequence, (Sect. 3.3.3).
The sequence under discussion is

$$x_{n+1} = 1 - x_n^2, \quad x_1 = a \in]0, 1[, \quad n \in \mathbb{N}^*.$$

This code uses the idea of a list more clearly.

```
With[{m=7},   (* Number of terms *)
u[1]={.25,0.9,(-1+Sqrt[5])/2};
u[n_]:=1-u[n-1]^2;
Table[data@i=Table[{n,u[n][[i]]},{n,m}],{i,3}];
{Grid[Table[{k,u[k][[1]],u[k][[2]],Simplify[u[k][[3]]]},{k,m}],
Alignment→Left,Frame→All,
Background→{{{None,RGBColor[0,0,1,.4],
RGBColor[1,0,0,.5],RGBColor[0,1,0,.5]}},None}],
ListLinePlot[Table[data@i,{i,3}],
PlotStyle→{Blue,Red,Darker[Green]},
Mesh→Full,MeshStyle→{Black,PointSize[Medium]},
Ticks→
{Table[μ,{μ,m}],{{.25,".25"},{−1+√5/2,"−1+√5/2"}{.9,".9"},1}},
ImageSize → {280,200}]}]]}
]
```

See Fig. 3.4 below.

1	0.25	0.9	$\frac{1}{2}\left(-1+\sqrt{5}\right)$
2	0.9375	0.19	$\frac{1}{2}\left(-1+\sqrt{5}\right)$
3	0.121094	0.9639	$\frac{1}{2}\left(-1+\sqrt{5}\right)$
4	0.985336	0.0708968	$\frac{1}{2}\left(-1+\sqrt{5}\right)$
5	0.0291124	0.994974	$\frac{1}{2}\left(-1+\sqrt{5}\right)$
6	0.999152	0.0100274	$\frac{1}{2}\left(-1+\sqrt{5}\right)$
7	0.00169434	0.999899	$\frac{1}{2}\left(-1+\sqrt{5}\right)$

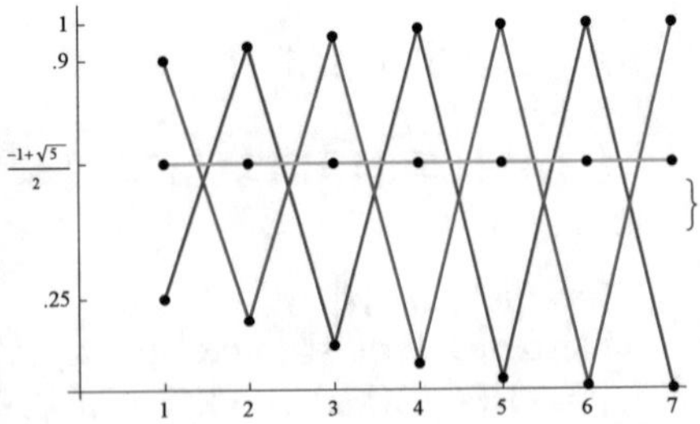

Fig. 3.4 Again the third sequence

Chapter 4
Sorting Algorithms

4.1 Introduction

An important problem of theoretical computer science is the *sorting* problem. The statement of the problem (in a very simple but satisfactory form) is: given a set of *n* real numbers, find an algorithm that sorts this set of entries in a minimum number of steps (comparisons and interchanges). We mention a fundamental work on this topic by D. Knuth [38].

We have already seen that Sort is a built-in Wolfram language function. Some examples are introduced below:

```
Clear[a,b,c,d,x,y]
Sort[{d,b,c,a}],    (* Sort implicitly provides a list in ascending order *)
Sort[{4,1,3,2,2},Greater],    (* Here we ask to sort in descending
order *)
Sort[{{a,2},{c,1},{d,3}},#1[[2]]< #2[[2]]&],    (* Sort also supports
pairs of numbers. A criterion is given after the list of objects to be
sorted *)
Sort[y,x²,x+y,y³],
Sort[{ "cat" , "fish" , "catfish" , "Cat" }],    (* Sort supports text *)
Sort[{ "aa" , "abb" , "ba" ,b, "aaa" }],
Sort[{Pi,E,2,3,1,1-I,Sqrt@2}],    (* Complex numbers are present *)
Sort[{-11,10,2,1,-4, 1-3I},Abs@#1<Abs@#2&]
{{a,b,c,d},{4,3,2,2,1},{{c,1},{a,2},{d,3}},{x²,y,y³,x+y},
{cat,Cat,catfish,fish},{aa,aaa,abb,b,ba},{1, 1 − i, 2, 3, √2, e, π},
{1,2,1-3 i,-4,10,-11}}
```

© Springer International Publishing AG 2017
M. Mureşan, *Introduction to Mathematica® with Applications*,
DOI 10.1007/978-3-319-52003-2_4

Sort@Join[Log@Range@5,Sqrt@Range@5]   (* The sorting is
performed alphanumerically; Log[2] is considered a symbol but a
number *)
Sort@N@Join[Log@Range@5,Sqrt@Range@5]   (* Numerical it
worked properly *)
Sort[Join[Log@Range@5,Sqrt@Range@5],Less]   (* Adding an
ordering relation, Sort works properly *)

$$\left\{0, 1, 2, \sqrt{2}, \sqrt{3}, \sqrt{5}, \text{Log}[2], \text{Log}[3], \text{Log}[4], \text{Log}[5]\right\}$$

$$\{0., 0.693147, 1., 1.09861, 1.38629, 1.41421, 1.60944, 1.73205, 2., 2.23607\}$$

$$\left\{0, \text{Log}[2], 1, \text{Log}[3], \text{Log}[4], \sqrt{2}, \text{Log}[5], \sqrt{3}, 2, \sqrt{5}\right\}$$

RandomInteger[{-5,5},{10,3}];   (* A list of ten triples whose
components are random integers in [-5,5] *)
Sort[%,Norm@#1<Norm@#2&]   (* It sorts the previous list
ascending with respect to their norms *)

$$\{\{-3, 0, 3\}, \{1, 4, 2\}, \{4, 2, 1\}, \{1, -4, -3\}, \{3, -1, 4\}, \{2, 3, -5\}, \{-1, 5, 4\},$$
$$\{5, -4, 4\}, \{-4, -5, -5\}, \{5, -5, 5\}\}$$

Sort[{I,1+I,1-I,2+3 I},Re@#1<Re@#2&]   (* It sorts ascending with
respect to the real parts *)

{i,1-i,1+i,2+3 i}

Sort[{4,1,3,2},(Print@{#1,#2};#1>#2)&]   (* The comparisons
are printed step by step *)

$\{4, 1\}$
$\{3, 2\}$
$\{1, 3\}$
$\{4, 3\}$
$\{1, 2\}$
$\{4, 3, 2, 1\}$

4.2 Sorting Methods

We have mentioned that *Mathematica* has a built-it Wolfram language symbol for
the sorting problem, that is, Sort. It was used earlier. At the same time each one can
imagine his/her own sorting algorithm. Here we review five sorting methods.

4.2.1 Selection Sort

First, we find the smallest element in the input list and interchange it with the first
element. Then we find the next smallest element and interchange it with the second
element. Continue in this way until the entire sequence is sorted.

In other words we say that the *selection sort* divides the input list into two parts: the sublist of items already sorted, which is built up from left to right located at the left-hand side of the list (input), and the sublist of items remaining to be sorted that occupy the rest of the list. Initially, the sorted sublist is null, and the unsorted sublist is the entire input list.

Mathematica has its own selection sort. It requires the "Combinatorica'" package.

```
Needs[" Combinatorica"]
RandomInteger[{1,20},15]
SelectionSort[%,Less]
{3, 12, 16, 1, 3, 18, 3, 1, 7, 16, 5, 4, 19, 3, 17}
{1, 1, 3, 3, 3, 3, 4, 5, 7, 12, 16, 16, 17, 18, 19}
```

A code for selection sort follows. The elements which will be interchanged are given in blue.

```
list=RandomSample[Range@10,7];
Print[" initial list ",list]    (* Initial input list *)
Do[If[list[[i]]>list[[j]],plist=list;
plist[[i]]=Style[plist[[i]],Blue,Bold,16];
plist[[j]]=Style[plist[[j]],Blue,Bold,16];
Print@plist;
list[[{i,j}]]=list[[{j,i}]]]],{i,Length@list},{j,i+1,Length@list}];
Print[" sorted list ",list]    (* Sorted output list *)
initial list {7, 6, 4, 3, 9, 1, 5}
{7,6,4,3,9,1,5}
{6,7,4,3,9,1,5}
{4,7,6,3,9,1,5}
{3,7,6,4,9,1,5}
{1,7,6,4,9,3,5}
{1,6,7,4,9,3,5}
{1,4,7,6,9,3,5}
{1,3,7,6,9,4,5}
{1,3,6,7,9,4,5}
{1,3,4,7,9,6,5}
{1,3,4,6,9,7,5}
{1,3,4,5,9,7,6}
{1,3,4,5,7,9,6}
{1,3,4,5,6,9,7}
sorted list {1,3,4,5,6,7,9}
```

Another short code for selection sort algorithm was introduced in [53, Chap. 10].

```
selectionSort[a_List]:=Flatten[{Min@a,
If[Length@a>1,selectionSort@Drop[a,First@Position[a,Min@a]],]}];
```

We test it.

```
RandomInteger[{1,9},15]
selectionSort@%
```
$\{6, 2, 2, 7, 2, 3, 8, 5, 7, 8, 2, 2, 4, 6, 5\}$
$\{2, 2, 2, 2, 2, 3, 4, 5, 5, 6, 6, 7, 7, 8, 8\}$

We introduce below a short code for the selection sort algorithm. The feature of this code is that the keys to be interchanged are written in augmented bold type, whereas the keys compared are given in augmented italic type.

```
appendto[char_,ii_,jj_]:=Block[{c},
c=a;
c[[ii]]=Style[a[[ii]],char,Blue,16];
c[[jj]]=Style[a[[jj]],char,Blue,16];
AppendTo[aa,c]];

With[{m=10,n=10},     (* For a random selection of n terms in
Range[m] *)
a=RandomSample[Range@m,n]];
Print[" Input (unsorted) list : ",%]
aa={{%%}};
Do[min=a[[i]];minj=i;
Do[If[a[[j]]<min,appendto[Italic,minj,j];
min=a[[j]];minj=j],{j,i+1,Length@a}];
If[i≠minj,appendto[Bold,i,minj]];
a[[{i,minj}]]=a[[{minj,i}]],{i,Length@a-1}
];
AppendTo[aa,a];
aa
Print[" Output (sorted) list : ",a]
```

The sorting steps are presented in Fig. 4.1.

4.2.2   Insertion Sort

Let $a = \{a_1, \ldots, a_n\}$ be the list to be sorted. At the beginning and after each iteration of the algorithm, the list consists of two parts: the first part $a_1, \ldots, a_{i-1}$ is already sorted, and the second part $a_i, \ldots, a_n$ is still unsorted $i \in \{1, \ldots, n\}$. In order to insert element a_i into the sorted part, it is compared with a_{i-1}, a_{i-2}, etc. When an element a_j with $a_j \leq a_i$ is found, a_i is inserted behind it. If no such element is found, then a_i is inserted at the beginning of the sequence. After inserting the element a_i, the length of the sorted part has increased by one. In the next iteration, a_{i+1} is inserted into the sorted part. While at the beginning the sorted part consists of element a_1 only, at the end it consists of all the elements $a_1, \ldots, a_n$. This is the *insertion sort* algorithm.

```
Input (unsorted) list : {2, 4, 9, 7, 1, 3, 10, 6, 8, 5}
```

$\{\{2, 4, 9, 7, 1, 3, 10, 6, 8, 5\},$
$\{2, 4, 9, 7, 1, 3, 10, 6, 8, 5\}, \{2, 4, 9, 7, 1, 3, 10, 6, 8, 5\},$
$\{1, 4, 9, 7, 2, 3, 10, 6, 8, 5\}, \{1, 4, 9, 7, 2, 3, 10, 6, 8, 5\},$
$\{1, 2, 9, 7, 4, 3, 10, 6, 8, 5\}, \{1, 2, 9, 7, 4, 3, 10, 6, 8, 5\},$
$\{1, 2, 9, 7, 4, 3, 10, 6, 8, 5\}, \{1, 2, 9, 7, 4, 3, 10, 6, 8, 5\},$
$\{1, 2, 3, 7, 4, 9, 10, 6, 8, 5\}, \{1, 2, 3, 7, 4, 9, 10, 6, 8, 5\},$
$\{1, 2, 3, 4, 7, 9, 10, 6, 8, 5\}, \{1, 2, 3, 4, 7, 9, 10, 6, 8, 5\},$
$\{1, 2, 3, 4, 7, 9, 10, 6, 8, 5\}, \{1, 2, 3, 4, 5, 9, 10, 6, 8, 7\},$
$\{1, 2, 3, 4, 5, 9, 10, 6, 8, 7\}, \{1, 2, 3, 4, 5, 6, 10, 9, 8, 7\},$
$\{1, 2, 3, 4, 5, 6, 10, 9, 8, 7\}, \{1, 2, 3, 4, 5, 6, 10, 9, 8, 7\},$
$\{1, 2, 3, 4, 5, 6, 10, 9, 8, 7\}, \{1, 2, 3, 4, 5, 6, 7, 9, 8, 10\},$
$\{1, 2, 3, 4, 5, 6, 7, 9, 8, 10\}, \{1, 2, 3, 4, 5, 6, 7, 8, 9, 10\}\}$

```
Output (sorted) list : {1, 2, 3, 4, 5, 6, 7, 8, 9, 10}
```

Fig. 4.1 Selection sort algorithm

We introduce below a short code in *Mathematica* for the insertion sort algorithm in [53, Chap. 10]. The keys to be moved are written in italic type, whereas the key to be inserted is given in bold blue type:

```
insertionSort[list_List]:=Module[{term,a=list},   (* Declaring it as
module, the code acts as a function that can be called several times *)
For[i=2,i≤Length@a,i++,term=a[[i]];j=i-1;
While[j≥1&&a[[j]]>term,
b=a;b[[i]]=Style[a[[i]],Italic,Blue,16];
b[[j]]=Style[a[[j]],Italic,Blue,16];
a[[j+1]]=a[[j]];
AppendTo[aa,b];j--];
a[[j+1]]=term;b=a;
b[[j+1]]=Style[a[[j+1]],Bold,Blue,16];
AppendTo[aa,b]];
AppendTo[aa,a]];

aa={{5,2,4,6,1}};   (* Here is the input (unsorted) list; the list aa
will contain all the steps of the algorithm *)
insertionSort@Flatten[aa,1]   (* The last element of the output
consists in the corresponding sorted list *)
```

The sorting steps are presented in Fig. 4.2.

We can visualize the insertion sort algorithm by converting the successive lists into permutation plots with animation.

$\{\{5, 2, 4, 6, 1\}, \{5, 2, 4, 6, 1\}, \{2, 5, 4, 6, 1\}, \{2, 5, 4, 6, 1\},$
$\{2, 4, 5, 6, 1\}, \{2, 4, 5, 6, 1\}, \{2, 4, 5, 6, 1\}, \{2, 4, 5, 6, 6\},$
$\{2, 4, 5, 5, 6\}, \{2, 4, 4, 5, 6\}, \{1, 2, 4, 5, 6\}, \{1, 2, 4, 5, 6\}\}$

Fig. 4.2 Insertion sort algorithm

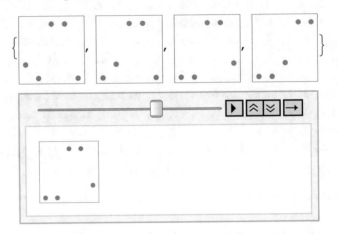

Fig. 4.3 Insertion sort with permutations and animation

```
permutationPlot[list_List]:=ListPlot[list,
PlotRange→{{.5,Length@list+.5},{.6,Length@list+.6}},
PlotStyle→PointSize[.4/Length@list],Axes→None,
FrameTicks→None,Frame→True,AspectRatio→1,
ImageSize→3.2Length@list(Length@list-1)]
With[{m=5,n=5},
a=RandomInteger[{1,m},n];
Print["Input (unsorted) sequence : ",a];
aa={a};
If[n>1,
Do[min=a[[i]];minj=i;appendto=0;
Do[If[a[[j]]<min,min=a[[j]];minj=j;appendto=1],{j,i+1,n}];
If[appendto==1,a[[{i,minj}]]=a[[{minj,i}]]; AppendTo[aa,a]],{i,n-1}
];
]]
Print[aa]
permutationPlot[#]&/@aa
ListAnimate[%] {2, 1, 5, 5, 1}
Input (unsorted) sequence : {2, 1, 5, 5, 1}
{{2, 1, 5, 5, 1}}
{{2, 1, 5, 5, 1}, {1, 2, 5, 5, 1}, {1, 1, 5, 5, 2}, {1, 1, 2, 5, 5}}
```

The sorting steps are presented in Fig. 4.3.

4.2.3 Mergesort

Mergesort algorithm is based on two steps, [53, Chap. 10]. The first one splits the input list into smaller groups by halving it until the groups had only one or two elements. Then it merges the groups back together so that their elements are in order. This is an algorithm of the "divide and conquer" type. Below we introduce a code for mergesort.

```
merge[left_List,right_List]:=
Module[{lindex=1,rindex=1},
Table[Which[lindex>Length@left,right[[rindex++]],
rindex>Length@right,left[[lindex++]],left[[lindex]]≤right[[rindex]],
left[[lindex++]],True,right[[rindex++]]],{Length@left+Length@right}]]

mergeSort[m_List]:=Module[{middle},If[Length@m≥2,
middle=Ceiling[Length@m/2];
Apply[merge,Map[mergeSort,Partition[m,middle,middle,{1,1},{}]]],m]]

a=RandomInteger[{1,13},15]
mergeSort@a
{8, 8, 7, 1, 5, 4, 4, 5, 1, 1, 7, 8, 6, 2, 12}
{1, 1, 1, 2, 4, 4, 5, 5, 6, 7, 7, 8, 8, 8, 12}
```

4.2.4 Heapsort

We mention that there exists a built-in function heapsort which performs sorting using this algorithm.

```
Needs["Combinatorica"]
HeapSort[{9,5,1,8,6,4,3,7,2}]
{1, 2, 3, 4, 5, 6, 7, 8, 9}
```

The heapsort code introduced below is based on the version of Floyd's algorithm published in [26] and the corresponding *Mathematica* code given by [43, p. 99].

A list $a = \{a_1, \ldots, a_n\}$ is said to be a *heap* if $a_{\lfloor i/2 \rfloor} \leq a_i$, for all $i \in \{2, \ldots, n\}$. We present the heapsort algorithm in full detail.

○ Heapsort is a sorting algorithm of class $O(n \ln(n))$.
○ It is also known as sorting by the "method of heaps."
○ It is not recursive but is comparable to the quicksort.
○ Heapsort is a sorting algorithm "in situ," that is, it does not require additional structures; the sorting is performed in the space of the input list. We mention that there are many heapsort implementations, but it is not our aim to review them.

o Heapsort is similar to some extent to the selection sort.
o At each step, the smallest element is selected, moved behind the input list, and
   ignored along the next steps. Thus, the amount of work is gradually reduced.
o In order to find the smallest element of the input, it is transformed into a heap.

Each input list a in the next grid

    Clear[a];
    headerheap=Join[{"index"},Table[k,{k,9}]];
    list=Join[{a},Table[a_k,{k,9}]];
    Grid[Join[{headerheap},{list}],Frame→All,Alignment→Center,
    FrameStyle→Thin,Background→{None,{Green,Yellow}}]

can be transformed into a binary tree as the figure below shows.
   See Fig. 4.4.

    TreePlot[{$a_1 \to a_2, a_1 \to a_3, a_2 \to a_4, a_2 \to a_5, a_3 \to a_6, a_3 \to a_7$,
    $a_4 \to a_8, a_4 \to a_9$},Automatic,VertexLabeling→True,
    PlotStyle→Black,ImageSize→50]

See Fig. 4.5.
   A *binary tree* is a tree-like structure that is rooted, and each vertex has at most
two children, and each child of a vertex is designed as its left or right child. There
is a vast literature on binary trees; we only mention [38, pp. 145–146].

o We note that the binary tree just obtained has the property that the index of each
   element is greater or equal to the index of its parent.
o a_1 is the root of the tree, so it has to be the minimal element of the input.
o In a heap the minimal element is located on the top of it.
o If one of the conditions $a_i \le a_{2i}$ or $a_i \le a_{2i+1}$ is not satisfied, then we interchange
   a_i with min $\{a_{2i}, a_{2i+1}\}$.
o In order to increase the speed of the algorithm, we perform the previous step from
   right to left.

Fig. 4.4 The list a

index	1	2	3	4	5	6	7	8	9
a	a_1	a_2	a_3	a_4	a_5	a_6	a_7	a_8	a_9

Fig. 4.5 The binary tree of a

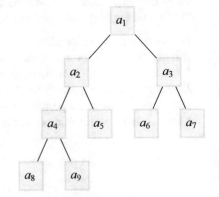

○ When the input is transformed step by step into a heap, the smallest element is at the top. This is interchanged with the last one, that is, a_n, which is at the top now, and we put it behind the smallest one, reducing by one the length of the new input.

We consider an example and perform the heapsort algorithm. Let $a = (9, 5, 1, 8, 6, 4, 3, 7, 2)$ be the input sequence (list) to be sorted. The binary tree representation of a is given by the Fig. 4.6.

TreePlot[{9 → 5, 9 → 1, 5 → 8, 5 → 6, 1 → 4, 1 → 3, 8 → 7, 8 → 2},
Automatic,VertexLabeling→True,PlotStyle→Black,ImageSize→150]

The sorting steps start as follow.

▷ We interchange $8 \longleftrightarrow 2$.
▷ We interchange $5 \longleftrightarrow 2$.
▷ We interchange $9 \longleftrightarrow \min\{2, 1\}$.

The steps introduced above are visualized below.

TreePlot[#,Automatic,9,VertexLabeling→True,PlotStyle→Black,
ImageSize→150]&/@
{{9 → 5, 9 → 1, 5 → 8, 5 → 6, 1 → 4, 1 → 3, 8 → 7, 8 → 2, 2 → 8},
{9 → 5, 9 → 1, 5 → 2, 5 → 6, 1 → 4, 1 → 3, 2 → 7, 2 → 8, 2 → 5},
{9 → 2, 9 → 1, 2 → 5, 2 → 6, 1 → 4, 1 → 3, 5 → 7, 5 → 8, 1 → 9}}

The first steps are exhibited in Fig. 4.7.
Next we go on as follows.

▷ We interchange $9 \longleftrightarrow \min\{4, 3\}$. Such interchanges will go on until 9 will arrive on the last row, and thus we get a heap. The smallest element lies on top of the binary tree.
▷ We interchange 1 with the last element of the last row, that is, $1 \longleftrightarrow 8$.
▷ We remove 1 and the link between 1 and 5.

Now we are in the situation introduced in the figure below.

Fig. 4.6 The binary tree of new a

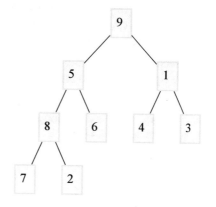

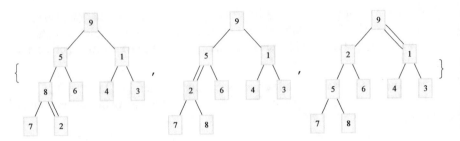

Fig. 4.7 The first steps of the heapsort of a

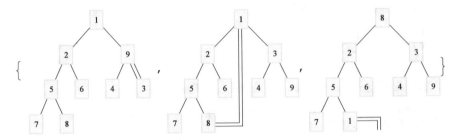

Fig. 4.8 The second steps of the heapsort of a

{TreePlot[{1 → 2, 1 → 9, 2 → 5, 2 → 6, 5 → 7, 5 → 8, 9 → 4, 9 → 3, 3 → 9},
Automatic,1,VertexLabeling→True,PlotStyle→Black,ImageSize→150],
Show[Graphics[{{PointSize[Medium],Point[{0,0}]},
Line[{{{1,0},{1.6,0},{1.6,2.6}},{{1,-.08},{1.68,-.08},{1.68,2.6}}}]}]],
TreePlot[{1 → 2, 1 → 3, 2 → 5, 2 → 6, 5 → 7, 5 → 8, 3 → 4, 3 → 9},
Automatic,1,VertexLabeling→True,PlotStyle→Black],PlotRange→All,
ImageSize→145],
Show[Graphics[{{PointSize[Medium],Point[{0,0}]},
Line[{{{1,0.1},{1.6,0.1},{1.6,-.2}},{{1,.02},{1.52,.02},{1.52,-.2}}}]}]],
TreePlot[{8 → 2, 8 → 3, 2 → 5, 2 → 6, 5 → 7, 5 → 1, 3 → 4, 3 → 9},
Automatic,8,VertexLabeling→True,PlotStyle→Black],PlotRange→All,
ImageSize→145]}

The second steps are exhibited in Fig. 4.8.

We further work with the remained binary tree whose length is decreased by 1.

▷ Due to the previous steps, the vertices 5, 3 and 2 satisfy the condition $a_{\lfloor i/2 \rfloor} \leq a_i$.
   Furthermore, this condition is also satisfied by the vertices 8 and 2. Therefore,
   we interchange 8 ⟷ 2.
▷ We interchange 8 ⟷ 5.
▷ We interchange 8 ⟷ 7.

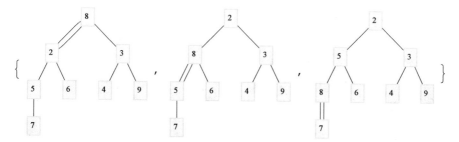

Fig. 4.9 The third steps of the heapsort of *a*

The steps that we made can be visualized below.

{TreePlot[{8 → 2, 8 → 3, 2 → 5, 2 → 6, 5 → 7, 3 → 4, 3 → 9, 2 → 8},
Automatic,8, VertexLabeling → True, PlotStyle → Black,
ImageSize → 150],
TreePlot[{2 → 8, 2 → 3, 8 → 5, 8 → 6, 5 → 7, 3 → 4, 3 → 9, 5 → 8},
Automatic, 2, VertexLabeling → True, PlotStyle → Black,
ImageSize → 150],
TreePlot[{2 → 5, 2 → 3, 5 → 8, 8 → 7, 5 → 6, 3 → 4, 3 → 9, 7 → 8},
Automatic, 2, VertexLabeling → True, PlotStyle → Black,
ImageSize → 150]}

The third steps are exhibited in Fig. 4.9.

The last tree is a heap; thus, we proceed in the following way:

▷ We interchange 2 ⟷ 8.
▷ We remove 2 from the tree and the link between 2 and 7.
▷ The tree just obtained is ready for a new step of the algorithm.

These steps are shown in the figure below.

{Show[Graphics[{Line[{{{0,.058},{1.18,.058},{1.18,2.6}},
{{0,-.05},{1.27,-.05},{1.27,2.6}}}]}],
TreePlot[{2 → 5, 2 → 3, 5 → 7, 7 → 8, 5 → 6, 3 → 4, 3 → 9},
Automatic,2,VertexLabeling→True,PlotStyle→Black],
ImageSize→145]}
Show[Graphics[{Line[{{{0,.058},{.65,.058},{.65,-.15}},
{{0,-.03},{.55,-.03},{.55,-.15}}}]}]],
TreePlot[{8 → 5, 8 → 3, 5 → 7, 5 → 6, 3 → 4, 3 → 9, 7 → 2},Automatic,8,
VertexLabeling→True,PlotStyle→Black],ImageSize→145],
TreePlot[{8 → 5, 8 → 3, 5 → 7, 5 → 6, 3 → 4, 3 → 9}, Automatic,8,
VertexLabeling→True, PlotStyle→Black,ImageSize→145]}

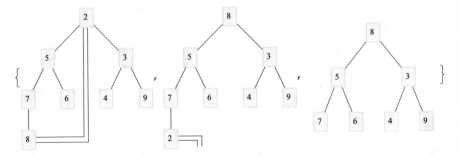

Fig. 4.10 The fourth steps of the heapsort of a

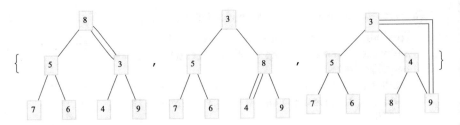

Fig. 4.11 The fifth steps of the heapsort of a

The fourth steps are exhibited in Fig. 4.10.
We continue the algorithm.

▷ We interchange $8 \longleftrightarrow \min\{5, 3\}$.
▷ We interchange $8 \longleftrightarrow 4$.
▷ We have a heap and interchange $9 \longleftrightarrow 3$.

These steps are visualized in the next figure.

> {TreePlot[{8 → 5, 8 → 3, 5 → 7, 5 → 6, 3 → 4, 3 → 9, 3 → 8},
> Automatic,8,VertexLabeling→True,PlotStyle→Black,ImageSize→145],
> TreePlot[{3 → 5, 5 → 7, 5 → 6, 3 → 8, 8 → 4, 8 → 9, 4 → 8},
> Automatic,3,VertexLabeling→True,PlotStyle→Black, ImageSize→145],
> Show[Graphics[{Line[{{{2.4,.058},{2.4,2-.4},{1+.2,2-.4}},
> {{2.48,.058},{2.48,2-.32},{1+.2,2-.32}}}]}],
> TreePlot[{3 → 5, 5 → 7, 5 → 6, 3 → 4, 4 → 8, 4 → 9},Automatic,3,
> VertexLabeling→True,PlotStyle→Black],ImageSize→145]}

The fifth steps are exhibited in Fig. 4.11.
We go on to the next steps.

▷ We remove 3 from the tree and the link between 3 and 4.
▷ We interchange $9 \longleftrightarrow 4$.
▷ We interchange $9 \longleftrightarrow 8$ and get a heap.

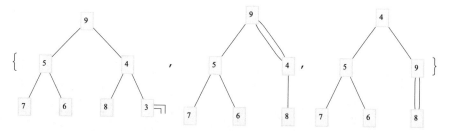

Fig. 4.12 The sixth steps of the heapsort of a

These steps are visualized below.

> {Show[Graphics[{Line[{{{2.5,.058},{2.8,.058},{2.8,-.1}},
> {{2.5,-.01},{2.72,-.01},{2.72,-.1}}}]}]],
> TreePlot[{9 → 5, 9 → 4, 5 → 7, 5 → 6, 4 → 8, 4 → 3},Automatic,9,
> VertexLabeling→True,PlotStyle→Black,ImageSize→145]],
> TreePlot[{9 → 5, 9 → 4, 5 → 7, 5 → 6, 4 → 8, 4 → 9},Automatic,9,
> VertexLabeling→True,PlotStyle→Black, ImageSize→145],
> TreePlot[{4 → 5, 4 → 9, 5 → 7, 5 → 6, 9 → 8, 8 → 9},Automatic,4,
> VertexLabeling→True,PlotStyle→Black,ImageSize→145]}

The sixth steps are exhibited in Fig. 4.12.
  We go on to the next steps.

▷ We interchange 9 ⟷ 4.
▷ We remove 4.
▷ We interchange 9 ⟷ 5 and get a heap.

These steps are visualized in the figure below.

> {Show[Graphics[Line[{{{1.75,.058},{1.09,0.058},{1.09,1.62}},
> {{1.75,.01},{1.05,.01},{1.05,1.62}}}]}],
> TreePlot[{4 → 5, 4 → 8, 5 → 7, 5 → 6, 8 → 9},Automatic,4,
> VertexLabeling→True, PlotStyle→Black],ImageSize→130],
> Show[Graphics[{Line[{{{1.75,.058},{2.15,0.058},{2.15,-.15}},
> {{1.75,-.03},{2.08,-.03},{2.08,-.15}}}]}]],
> TreePlot[{9 → 5, 9 → 8, 5 → 7, 5 → 6, 8 → 4},Automatic,9,
> VertexLabeling→True, PlotStyle→Black],ImageSize→150],
> TreePlot[{9 → 5, 9 → 8, 5 → 7, 5 → 6, 5 → 9},Automatic,9,
> VertexLabeling→True,PlotStyle→Black,ImageSize→130]}

The seventh steps are exhibited in Fig. 4.13.
  We continue to the next steps.

▷ We interchange 9 ⟷ 5.
▷ We interchange 9 ⟷ 6 and get a heap.
▷ We interchange 5 ⟷ 9 and remove 5 and the link between 5 and 9.
▷ We interchange 9 ⟷ 6.

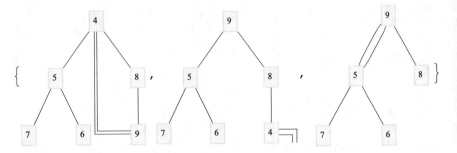

Fig. 4.13 The seventh steps of the heapsort of *a*

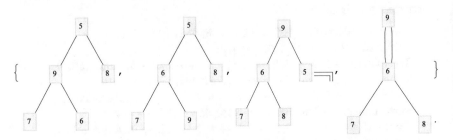

Fig. 4.14 The eighth steps of the heapsort of *a*

The visualizations of these steps follow.

> {TreePlot[{5 → 9, 5 → 8, 9 → 7, 9 → 6},Automatic,5,
> VertexLabeling→True,PlotStyle→Black,ImageSize→120],
> TreePlot[{5 → 6, 5 → 8, 6 → 7, 6 → 9},Automatic,5,
> VertexLabeling→True,PlotStyle→Black,ImageSize→120],
> Show[Graphics[{Line[{{{1.52,.830},{1.85,.830},{1.85,.72}},
> {{1.52,.9},{1.908,.9},{1.908,.72}}}]}]],
> TreePlot[{9 → 6, 9 → 8, 6 → 7, 6 → 5},Automatic,9,
> VertexLabeling→True, PlotStyle→Black,ImageSize→120],
> TreePlot[{9 → 6, 9 → 5, 6 → 7, 6 → 8},Automatic,9,
> VertexLabeling→True, PlotStyle→Black],ImageSize→120],
> TreePlot[{9 → 6, 6 → 7, 6 → 8, 6 → 9},Automatic,9,
> VertexLabeling→True, PlotStyle→Black,ImageSize→110]}

The eighth steps are exhibited in Fig. 4.14.
    We continue to the next steps.

▷   We interchange 9 ⟷ 7 and get a heap.
▷   We interchange 9 ⟷ 6.
▷   We remove 6 from the tree and the link between 6 and 7.

Fig. 4.15 The ninth steps of
the heapsort of a

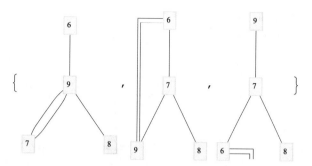

The visualizations of these steps follow.

{TreePlot[{6 → 9, 9 → 7, 9 → 8, 7 → 9},Automatic,6,
VertexLabeling→True,PlotStyle→Black,ImageSize→120],
Show[Graphics[{Line[{{{.04,0},{.04,2-.18},{.38,2-.18}},
{{.0,0},{.0,2-.12},{.38,2-.12}}}]}]],
TreePlot[{6 → 7, 7 → 9, 7 → 8},Automatic,6,VertexLabeling→True,
PlotStyle→Black],ImageSize→90],
Show[Graphics[{Line[{{{.04,0.04},{.44,0.04},{.44,-.1}},
{{.04,-0.03},{.4-.01,-0.03},{.4-.01,-.1}}}]}]],
TreePlot[{9 → 7, 7 → 6, 7 → 8},Automatic,9,VertexLabeling→True,
PlotStyle→Black],ImageSize→90]}

The ninth steps are exhibited in Fig. 4.15.
    We continue to the next steps.

▷ We interchange 9 ⟷ 7.
▷ We interchange 9 ⟷ 8 and get a heap.
▷ We interchange 9 ⟷ 7.
▷ We remove 7 from the tree and the link between 7 and 8.

These steps are visualized in the next figure.

{TreePlot[{9 → 7, 7 → 8, 7 → 9},Automatic,9,
VertexLabeling→True,PlotStyle→Black,ImageSize→{60,150}],
TreePlot[{7 → 9, 9 → 8, 8 → 9},Automatic,7,
VertexLabeling→True,PlotStyle→Black,ImageSize→{60,150}],
Show[Graphics[{Line[{{{.04,0.03},{.2,.03},{.2,2-.051},{.04,2-.051}},
{{.04,-0.03},{.3-.05,-.03},{.3-.05,2}, {.04,2}}}]}]]],
TreePlot[{7 → 8, 8 → 9},Automatic,7,VertexLabeling→True,
PlotStyle→Black],ImageSize→{60,150}]
Show[Graphics[{Line[{{{.04,0.03},{.27,.03},{.27,-.08}},
{{.04,-0.03}, {.3-.08,-.03},{.3-.08,-0.08}}}]}]],
TreePlot[{9 → 8, 8 → 7},Automatic,9,VertexLabeling→True,
PlotStyle→Black],ImageSize→{50,150}]}

Fig. 4.16 The tenth steps of the heapsort of a

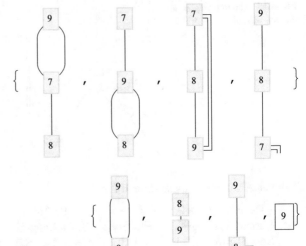

Fig. 4.17 The eleventh steps of the heapsort of a

The tenth steps are exhibited in Fig. 4.16.

We continue with the final steps.

▷ We interchange $9 \longleftrightarrow 8$ and get a heap.
▷ We interchange $9 \longleftrightarrow 8$ and remove 8 and the link between 8 and 9.
▷ The last and largest element remained, i.e., 9.

These final steps are visualized in the next figure.

```
{TreePlot[{9 → 8, 8 → 9},Automatic,9,
VertexLabeling→True,PlotStyle→Black,ImageSize→{40,80}],
Show[Graphics[{Point[{0,0}],Point[{0,1}]}],TreePlot[{8 → 9},
Automatic,8,VertexLabeling→True,PlotStyle→Black],
ImageSize→{50,80}],
Show[Graphics[{Line[{{{.04,0.03},{.27,.03},{.27,-.08}},
{{.04,-0.03},{.3-.08,-.03},{.3-.08,-0.08}}}]}],
TreePlot[{9 → 8},Automatic,9,VertexLabeling→True,
PlotStyle→Black],ImageSize→{40,80}],
Show[Graphics[
{Line[{{-.03,-0.04},{.03,-.04},{.03,.04},{-.03,0.04},{-.03,-0.04}}],
Text[9,{0,0}]}],ImageSize→{18,33}]}
```

The eleventh steps are exhibited in Fig. 4.17

The heapsort algorithm sorts the list $a = \{a_1, \ldots, a_n\}$, requiring no more than $2 (2^p - 2) (p - 1)$, or approximately $2n (\text{Log}_2[n] - 1)$ comparisons and half as many interchanges in the worst case to sort $n = 2^p - 1$ items. The algorithm is easier to follow if a is thought of as a tree, with $a_{\lfloor i/2 \rfloor}$ the father of a_i for $1 < i < n$.

Now a *Mathematica* code of the above introduced algorithm follows.

```
heaps={};   (* We collect the partial heaps *)
heapSort[p_List]:=Module[{heap=heapify@p,min},
Append[Table[min =First@heap;
heap[[1]]=heap[[n]];

heap=heapify[Drop[heap,-1],1];AppendTo[heaps,heap];
min,{n,Length@p,2,-1}],Max@heap]]/;Length@p>0
heapify[p_List]:=Module[{j,heap=p},Do[heap=heapify[heap,j],
{j,Quotient[Length@p,2],1,-1}];heap]
heapify[p_List,k_Integer]:=Module[{hp=p,i=k,m,n=Length@p},
While[(m=2 i)≤n,
If[m<n&&Less[hp[[m+1]],hp[[m]]],m++];
If[Less[hp[[m]],hp[[i]]],hp[[{i,m}]]=hp[[{m,i}]]];
i=m];
hp
]

a={9,5,1,8,6,4,3,7,2};
heapSort@a
Table[heaps[[i]],{i,1,Length@heaps}]   (* Here is the list of partial
heaps *)
{1,2,3,4,5,6,7,8,9}
{{2,5,3,7,6,4,9,8},{3,5,4,7,6,8,9},{4,5,8,7,6,9},{5,6,8,7,9},{6,7,8,9},
{7,9,8},{8,9},{9}}

trees={};
Do[a=heaps[[m]];
n=Length@a
If[n==0,tree={};Goto["print"],
If[n==1,tree=a;Goto["print"],
tree={};
If[n==2,AppendTo[tree,a[[1]]→a[[2]]];Goto["print"],
tree={};j=1;

Do[
j++;AppendTo[tree,a[[i]]→a[[j]]];
j++;If[j>n,Goto["print"]];
AppendTo[tree,a[[i]]→a[[j]]],{i,⌊n/2⌋}]
]]];
Label["print"];
AppendTo[trees,tree],
{m,Length@heaps-1}];
Table[TreePlot[trees[[k]],Automatic,trees[[k,1,1]],VertexLabeling→True,
ImageSize→ 10(15 − ⌊k/2⌋)],{k,Length@trees}]
```

The steps of the above algorithm are exhibited in Fig. 4.18.

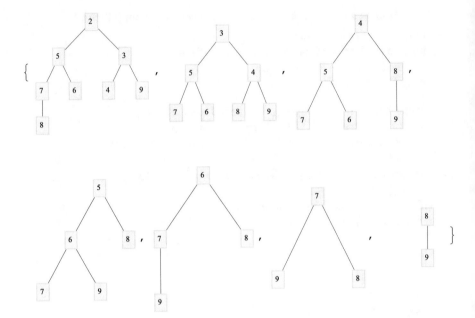

Fig. 4.18 The steps of the above heapsort algorithm for a

4.2.5   Quicksort

The *quicksort* algorithm was developed by C. A. R. Hoare and published in [34]. A series of interesting writings on this topic were published by R. Sedgewick. We mention here [61] and the reference therein. Some details may be found in [53, Chap. 10].

The quicksort algorithm starts by selecting an element, called the *pivot* point, from a list and then divides the list into two, with all elements less than the pivot element moved to or remained on its left and all elements greater than the pivot element moved to or remained on its right. The process is then continuously repeated in each sublist until the initial list is sorted, ascending in our case. The initial list is sorted when the length of each sublist is 1.

A short *Mathematica* code of the quicksort algorithm follows.

```
Clear[x]
quickSortcases[x_List]:=
Module[{pivot},
If[Length@x≤1,Return@x];   (* If the list is empty or it contains a
single element, the algorithm stops *)
pivot=RandomChoice@x;   (* It selects a random pivot from the list *)
Flatten@{   (* It flattens the three sorted lists *)
quickSortcases@Select[x,#<pivot&],   (* It selects all the
elements smaller than the pivot *)
```

```
Cases[x,y_/;y==pivot],
quickSortcases@Cases[x,y_/;y>pivot]}
]
```

```
With[{m=10,n=12},
list=RandomChoice[Range@m,n]]
quickSortcases@list
{5, 3, 3, 4, 2, 1, 4, 7, 8, 8, 4, 9}
{1, 2, 3, 3, 4, 4, 4, 5, 7, 8, 8, 9}
```

Another quicksort code is introduced below.

```
quickSortselect[{}]={};
quickSortselect[{x_,xs___}]:=Join[   (* It joins all the elements less
than, equal to, or greater than x *)
quickSortselect@Select[{xs},#≤x&],   (* It applies quickSortselect
to all picked elements in the list xs less than or equal to x *)
{x},
quickSortselect@Select[{xs},#>x&]   (* It applies quickSortselect to
all picked elements in the list xs greater than x *)
];
```

```
With[{m=10,n=12},
list=RandomChoice[Range@m,n]]
quickSortselect@list
{1, 5, 8, 7, 3, 6, 10, 4, 2, 7, 4, 9}
{1, 2, 3, 4, 4, 5, 6, 7, 7, 8, 9, 10}
```

Suppose we are interested in the time needed to sort a list.

Timing[*expr*] evaluates *expr* and returns a list of the time in seconds used, together with the result obtained. Then, one can use the next procedure:

```
With[{m=10,000,n=15,000},
list=RandomChoice[Range@m,n]];   (* The input list *)
Timing[quickSortcases@list;]
Timing[quickSortselect@list;]
{1.15625, Null}
{0.65625, Null}
```

Based on this simple test, we conclude that the quickSortselect code is clearly faster than the quickSortcases code.

Chapter 5
Functions

5.1 Definitions

We have already seen some built-in functions along the previous chapters. We can define our own functions depending on our needs or interests. Thus, we have the freedom to use plenty of mathematics. First, we consider some simple examples.

f[x_]:=Sin[x]+x²   (* A function is defined f(x) = sin x + x² *)
g[x_]:=f[f[x]+1]+2   (* A composition of a function g(x) = f(f(x) + 1) + 2 *)
{g[2],   (* The value of g at an integer; it is returned noncomputed *)
FullSimplify@g[2],   (* It is still noncomputed *)
TraditionalForm@g[2],
g[2.],   (* The value is given with five decimal digits *)
g[2]//N,
N[g[2],15],   (* The value is given with 15 decimals *)
N[#,15]&/@g[2]}   (* The same result as before *)
{2+(5+Sin[2])²+Sin[5+Sin[2]],2+(5+Sin[2])²+Sin[5+Sin[2]],
2+(5+sin(2))²+sin(5+sin(2)),36.5546,36.5546,36.5545586117012,
36.5545586117012}

It is not difficult to construct a function of several variables. An easy example is the code of finding the solutions of the second-degree polynomial with real coefficients.

© Springer International Publishing AG 2017
M. Mureşan, *Introduction to Mathematica® with Applications*,
DOI 10.1007/978-3-319-52003-2_5

Clear[a,b,c]   (* We want to avoid any interference of previous values of a, b, or c *)

rootsseconddegree[a_,b_,c_]:= $\left\{ \frac{-b-\sqrt{b^2-4a*c}}{2a}, \frac{-b+\sqrt{b^2-4a*c}}{2a} \right\}$;

{rootsseconddegree[1,2,1],   (* Now we test our formula for different sets of data *)

u=rootsseconddegree[1,2,-4],

rootsseconddegree[1,2,4]}

$\left\{ \{-1,-1\}, \left\{ \frac{1}{2}\left(-2-2\sqrt{5}\right), \frac{1}{2}\left(-2+2\sqrt{5}\right) \right\}, \right.$
$\left. \left\{ \frac{1}{2}\left(-2-2i\sqrt{3}\right), \frac{1}{2}\left(-2+2i\sqrt{3}\right) \right\} \right\}$

Suppose we need the second solution of the second equation, first under a simple form and then under a numerical form. We write

Simplify@u[[2]]

N@u[[2]]

$-1+\sqrt{5}$

1.23607

There exists a built-in function which can solve the second-degree equation easily. This is the built-in function Reduce.

Reduce[a x^2+b x+c==0,x]

$\left(a \neq 0 \&\& \left(x==\frac{-b-\sqrt{b^2-4a*c}}{2a} \| \frac{-b+\sqrt{b^2-4a*c}}{2a} \right) \right)$
$\| \left(a==0\&\&b \neq 0\&\&x== -\frac{c}{b} \right) \| (c==0\&\&b==0\&\&a==0)$

We can use this function for constrained equations.

Clear[x,y]

Reduce $\left[x^2 - y^2== 8, \{x,y\}, \text{Reals} \right]$

$\left(x \leq -2\sqrt{2}\&\& \left(y == -\sqrt{-8+x^2}\|y == \sqrt{-8+x^2} \right) \right) \|$
$\left(x \geq 2\sqrt{2}\&\& \left(y == -\sqrt{-8+x^2}\|y == \sqrt{-8+x^2} \right) \right)$

Clear[x,y,xy]

Reduce $\left[x^2-y^2== 8, \{x,y\}, \text{Integers} \right]$ (* We look for integer solutions *)

xy={x,y}//.{ToRules[%]}   (* The list of solutions *)

{x,y}=xy[[1]]   (* We extract the first pair of solutions *)

$(x == -3\&\&y == -1)\|(x == -3\&\&y == 1)\|$
$(x == 3\&\&y == -1)\|(x == 3\&\&y == 1)$

$\{\{-3,-1\}, \{-3,1\}, \{3,-1\}, \{3,1\}\}$

$\{-3,-1\}$

5.1.1 *Functions with Conditions*

We define the discontinuous function

$$f(x) = \begin{cases} -x, & \text{if } |x| < 1, \\ \sin x, & \text{if } 1 \le x < 2, \\ \cos x, & \text{otherwise.} \end{cases}$$

Using the built-in function Which (making the graph continuous), we write

```
f[x_]:=Which[Abs@x<1,        -x,
             1≤ Abs@x<2,     Sin@x,
             True,           Cos@x
             ];
Plot[f@x,{x,-3,3},ImageSize→200]
```
(* A graph is useful to see how *Mathematica* interprets the built-in function Which. The graph is continuous; the function is transformed into a set-valued function *)

```
f/@{0.5,Pi/2,Pi}
```
(* The values of the function f at some values of the argument *)

```
Limit[f@x,x→ 2]
```
(* 2 is a discontinuity point. The limit is considered side limit from the right-hand side *)

The graph of the above discontinuous function is exhibited by the built-in function Which in Fig. 5.1.

We note that if the function is discontinuous at certain points, its graph is continuous, and thus the function which represents that graph is a set-valued function or a multifunction.

Fig. 5.1 The graph and some values by the built-in function Which

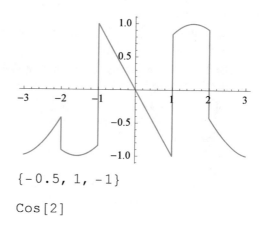

$$\{-0.5, 1, -1\}$$

```
Cos[2]
```

Fig. 5.2 The graph and some values by the built-in function Piecewise

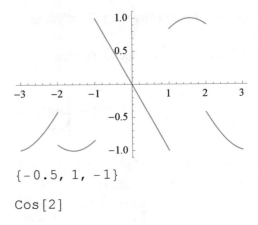

$\{-0.5, 1, -1\}$

Cos[2]

Using the built-in function Piecewise, the previous function looks like

g[x_]:=Piecewise[{{-x,Abs@x<1},{Sin@x,1≤Abs@x<2}},Cos@x]
Plot[g@x,{x,-3,3},ImageSize→200]
g/@{0.5,Pi/2,Pi}
Limit[g@x,x→2]

The graph of the above discontinuous function is exhibited by the built-in function Piecewise in Fig. 5.2.

We note that the built-in function Piecewise supplies the true graph of the function which in the present case has points of discontinuity.

The built-in function If supplies a continuous graph even if the function is discontinuous. Its output coincides with the one given by Which.

h[x_]:=If[Abs@x<1,-x,If[1≤Abs@x<2,Sin@x,Cos@x]];
Plot[h@x,{x,-3,3},ImageSize→200]
h/@{0.5,Pi/2,Pi}
Limit[h@x,x→2]   (* The limit is again considered side limit from the right-hand side *)

The graph of the above discontinuous function is exhibited by the built-in function If in Fig. 5.3.

p[x_]:=Piecewise[{{0,x<0},{1,x≥0}}]
Plot[p@x,{x,-1,1},PlotStyle→Thick,ImageSize→150,
Ticks→{{-1,1},{1}}]
p/@{-1,0,1}   (* At the discontinuity point x=0, it is returned the value 1 *)
Limit[p@x,x→0],Limit[p@x,x→0,Direction→-1],
Limit[p@x,x→0,Direction→1]   (* Because at x = 0 the function is discontinuous, the limit is considered side limit from the right-hand side *)

Another discontinuous function is exhibited by the built-in function Piecewise in Fig. 5.4.

Fig. 5.3 The graph and some values by the built-in function If

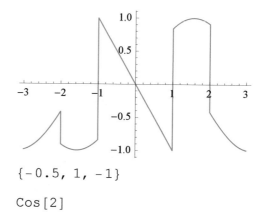

$\{-0.5, 1, -1\}$

$\texttt{Cos[2]}$

Fig. 5.4 Another graph and some values by the built-in function Piecewise

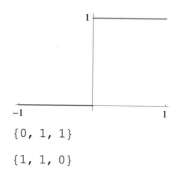

$\{0, 1, 1\}$

$\{1, 1, 0\}$

Fig. 5.5 The signum function

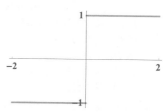

A similar discontinuous function is the signum (sign) function.

```
Plot[Sign@x,{x,-2,2},PlotStyle→{Darker[Green],Thick},
Ticks→{{-2,2},{-1,1}},TicksStyle→Directive[Red,Bold,12],
ImageSize→200]
```

The signum function is presented in Fig. 5.5 above.

Let us examine another example. In the following code appears /; and it is a shorthand of If.

```
Clear[f,g,h]
f[x_]:=Sqrt@x/;x>0
g[x_]:=Sqrt@x/;x<0
```

Fig. 5.6 Interpolation on
some finite and distinct data

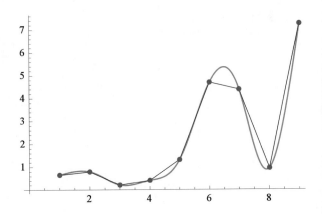

```
h[x_]:=Sqrt@x^2/;x<0
{f[4],f[-4],g[4],g[-4],h[2],h[-2]}
{2,f[-4],g[4],2 i,h[2],2}
```

Consider the same function defined by If. The two functions work identically for
$x > 0$. Otherwise, If returns Null.

```
g1[x_?Positive]:=Sqrt@x
g2[x_]:=If[x>0,Sqrt@x]
{{g1[4],g2[4]},{g1[-1],g2[-1]}}
{{2, 2}, {g1[−1], Null}}
```

Now an example of interpolation of some data follows.

```
data=Table[{i,RandomReal[i]},{i,9}]
ListLinePlot[data,Epilog→{Black,Line[data],PointSize[Medium],Red,
Point[data]},InterpolationOrder→3,ImageMargins→10,
ImageSize→250]
{{1, 0.652455}, {2, 0.794146}, {3, 0.207637}, {4, 0.40818}, {5, 1.3127},
{6, 4.69005}, {7, 4.38125}, {8, 0.94469}, {9, 7.26584}}
```

Below is presented an interpolation of some data in Fig. 5.6.

5.1.1.1   Collatz Conjecture

Collatz conjecture states that the algorithm consisting in substituting a given positive
integer n by Floor$[n/2]$ provided n is even and by $3n + 1$ provided n is odd reaches
eventually 1, [47, Chap. 1]. Formally we write

$$f(n) = \begin{cases} \lfloor n/2 \rfloor, & n \text{ is even,} \\ 3n + 1, & n \text{ is odd,} \end{cases}$$

and state that for any natural n, there exists a natural m such that f iterated m times at n reaches 1, i.e., $\underbrace{f(f(\ldots f(n)))}_{m \text{ times}} = 1$.

For example, starting with number 7, we write the Collatz successive iterations obtaining the following sequence:

$7 \longrightarrow 22 \longrightarrow 11 \longrightarrow 34 \longrightarrow 17 \longrightarrow 52 \longrightarrow 26 \longrightarrow 13 \longrightarrow 40 \longrightarrow 20 \longrightarrow 10 \longrightarrow 5 \longrightarrow 16 \longrightarrow 8 \longrightarrow 4 \longrightarrow 2 \longrightarrow 1$

A code for a step of this algorithm can be written as follows:

```
f[n_Integer]:=If[EvenQ[n],n/2,3n+1]
f[16]    (* Because 16 is even, we divide it by 2 *)
f[17]    (* Because 17 is odd, we multiply it by 3 and add 1 *)
8
52
```

We can write the above-defined function under the form:

```
g[n_Integer?EvenQ]:=n/2
g[n_]:=3n+1
g[8]
g[7]
4
22
```

Suppose we want to find out how many times one needs to apply g to numbers from 1 to 30 getting 1.

```
k=Length/@(NestWhileList[g,#,!#==1 &]&/@Range[30];)
m=Max@k    (* Find the largest number of steps needed to reach 1 *)
ListPlot[k,Filling→Axis,PlotStyle→Black,ImageSize→250,
PlotRange→All]
array=Table[{i,k[[i]]},{i,30}];
pointmaxat=Last[Sort[array,#1[[2]]< #2[[2]]&]][[1]]    (* We find the point
which requires the largest number of iteration to reach 1 *)
112
```

Some simple tests for Range[30] of the Collatz conjecture are presented in Fig. 5.7 below.

5.1.1.2 Square and Cubic Roots

Now we will introduce a very short code to compute $\sqrt{a}$, $a \geq 0$, using an approximate method.

```
Clear[f,a,ε]
f[a_, ε_]:= Module[{x = 1.,xp=a}, While[Abs[xp - x]≥ 10^−ε,xp = x;
x = (x+a/x)/2];x]
```

Fig. 5.7 Collatz conjecture for Range[30]

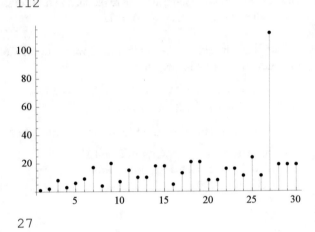

Suppose we want to find the square roots of some numbers with certain approximations. For $a = 3$ and $\varepsilon = 10^{-11}$, we write

N[f[3,11.],10]
1.73205

Another approach for $\sqrt{3}$, $a \geq 0$, follows

Nest[(# + a/#)/2&,1.0,5]/.a→3
1.73205

For the square root of a nonnegative number, we can use the built-in function $\sqrt{\Box}$. For example,

$N\left[\sqrt{3}, 11\right]$
1.7320508076

Now we introduce a very short code to compute, $\sqrt[3]{a}$, $a \geq 0$, using an approximate method.

```
Clear[f,a,ε]
f[a_,ε_]:=Module[{x = 1.,xp=a}, While[Abs[xp - x]≥ 10^{-ε},xp = x;
x = 1/3 (2x+ a /x² )];N[x]]
```

Suppose we want to find the cube root of a number with certain approximations, then we write

f[10,7.]
2.15443

We can use the built-in function $\sqrt[3]{\Box}$ to find $\sqrt[3]{10}$. Indeed

$N\left[\sqrt[3]{10}, 10\right]$
2.154434690

5.2 Differentiation and Integration

5.2.1 Differentiation of Functions

Suppose the function $f(x) = x^3 + 2x + 5$, $x \in \mathbb{R}$ is defined. Then, its first-order derivative is

```
Clear[f,x]
f[x_]:=x³+2x+5
f'@x
```
$2 + 3x^2$

or

```
D[f@x,x]
```
$2 + 3x^2$

To find the derivative at a given point, say 2.28, we write

```
%/.x→2.28
```
17.5952

or

```
f'@2.28
```
17.5952

For the second-order derivative, we write

```
f''@x
```
$6x$

or any of the following two

```
{Derivative[2][f][x],D[f[x],x,x]}
{6x,6x}
```

Immediately we have

```
{f'[π],Derivative[2][f][x]/.x→ π,D[f[x],x,x]/.x→ π}
{6π, 6π, 6π}
```

Suppose we have a function of three variables $f(x, y, z) = 2xy^2z^3 + \log(y^2 + 1) + y\cos(z^2)$, and we want to calculate $\frac{\partial^6 f(x,y,z)}{\partial x \partial y^2 \partial z^3}$. Then, we write

```
Clear[f,x,y,z]
f[x_,y_,z_]:=2xy²z³+Log[y²+1]+y Cos[z²];
{Derivative[1,2,3][f][x,y,z],
Derivative[1,1,0][f][x,y,z],   (* ∂²f(x,y,z)/∂x∂y *)
Derivative[1,1,0][f][x,y,z]/.y→1.25,
Derivative[1,1,0][f][x,y,z]/.{y→1.25,z→ 10⁻⁸}}
{24, 4yz³, 5.z³, 5.0000000000000005 *^ − 24}
```

Suppose we want to calculate Derivative[1,1,0][f][x,y,z]/.y→1.25 at z=1. Then, we write

```
%%/.z→1   (* It uses the last but one result *)
5.
```

5.2.2   *Integration of Functions*

We consider the following example:

```
Clear[f,h,x]
f[x_]:=x^4+1    (* f(x) = x^4 + 1 *)
h[x_]:=1/f[x]    (* h(x) = 1/f(x) *)
Integrate[h@x,x]    (* Indefinite integral, i.e., ∫ h(x) dx *)
Integrate[h@x,{x,-1,1}]    (* Definite integral, i.e., ∫_{-1}^{1} h(x) dx *)
TraditionalForm[%]
Simplify[%]
N[%]
```

$$\frac{-2\mathrm{ArcTan}\left[1-\sqrt{2}x\right]+2\mathrm{ArcTan}\left[1+\sqrt{2}x\right]-\mathrm{Log}\left[1-\sqrt{2}x+x^2\right]+\mathrm{Log}\left[1+\sqrt{2}x+x^2\right]}{4\sqrt{2}}$$

$$\frac{2\pi-\mathrm{Log}\left[17-12\sqrt{2}\right]}{4\sqrt{2}}$$

$$\frac{2\pi-\log\left(17-12\sqrt{2}\right)}{4\sqrt{2}}$$

$$\frac{2\pi-\mathrm{Log}\left[17-12\sqrt{2}\right]}{4\sqrt{2}}$$

1.73395

Sometimes apparently simple integrals lead to special functions.

```
Integrate[Sqrt[x]Sqrt[1+x],x]
FullSimplify[D[%,x]]
Integrate[Sqrt[x]Sqrt[1+x]Sqrt[2+x],x]
FullSimplify[D[%,x]]
```

$$\frac{\frac{1}{4}\left(\sqrt{x}\sqrt{1+x}(1+2x)-\mathrm{ArcSinh}\left[\sqrt{x}\right]\right)}{\sqrt{x}\sqrt{1+x}}$$

$$\frac{\frac{1}{5}\left(2\sqrt{x}(1+x)^{3/2}\sqrt{2+x}-\frac{1}{\sqrt{1+\frac{1}{x}}\sqrt{x}}\cdot\right.}{4\sqrt{\frac{1+x}{2+x}}\left(\sqrt{1+\frac{1}{x}}(2+x)+i\sqrt{x}\sqrt{\frac{2+x}{x}}\mathrm{EllipticE}\left[i\mathrm{ArcSinh}\left[\frac{1}{\sqrt{x}}\right],2\right]\right)}{\sqrt{x}\sqrt{1+x}\sqrt{2+x}}$$

Here EllipticE[ϕ,m] denotes the elliptic integral of the second kind. The incomplete elliptic integral of the second kind is defined

$$E(\phi,k) = \int_0^\phi \sqrt{1-k^2\mathrm{Sin}[\theta]}\,d\theta,$$

with $m = k^2$. For more details, visit the page mathworld.wolfram.com/EllipticIntegraloftheSecondKind.html.

We introduce some multiple integrals with an example.

Integrate[1/(x²+5),x,x,x]   (* The output contains the imaginary unit i *)
FullSimplify[%]   (* After changing to a simpler form, it looks more familiar *)
FullSimplify[[D[%,x,x,x]]   (* and we arrived to the starting function *)

$$\frac{1}{20}\left(10x + 2\sqrt{5}\left(-10 + x^2\right)\text{ArcTan}\left[\frac{x}{\sqrt{5}}\right] + 5i\sqrt{5}\text{Log}\left[\sqrt{5} - ix\right]\right.$$
$$\left.-5i\sqrt{5}\text{Log}\left[\sqrt{5} + ix\right] - 10x\text{Log}\left[5 + x^2\right]\right)$$
$$\frac{1}{10}\left(\sqrt{5}\left(-5 + x^2\right)\text{ArcTan}\left[\frac{x}{\sqrt{5}}\right] - 5x\left(-1 + \text{Log}\left[5 + x^2\right]\right)\right)$$
$$\frac{1}{5+x^2}$$

A multiple integral coincides with the iterative integral after simplifications if certain assumptions are satisfied [53, Chap. 8].

FullSimplify$\left[\text{Integrate}\left[1/(x^2 + 5), x, x, x\right]\right] ==$
FullSimplify$\left[\text{Integrate}\left[\text{Integrate}\left[\text{Integrate}\left[1/\left(x^2 + 5\right), x\right], x\right], x\right]\right]$
True

5.3   Functions and Their Graphs

We can integrate and/or differentiate an interpolating function as follows:

Needs["PlotLegends"]   (* We need the "PlotLegends" package *)
Clear[f,g,h,x]
f=Interpolation[{1,-2,5,-5}];
g[x_]=Integrate[f@x,x,x];
h[x_]=D[g@x,x,x];
Plot[{f@x,g@x,h@x},{x,1,4},Ticks→{{1,2,3,4},{-5,-2,1,5}},
AxesOrigin→{0,0},PlotLegend→{f,g,h},
LegendPosition→{1.1,-0.4},ImageSize→400]

The built-in function PlotLegends is presented in Fig. 5.8 below.

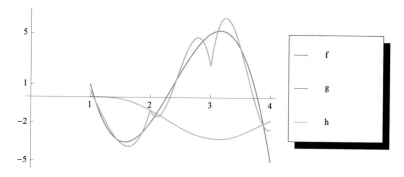

Fig. 5.8 The built-in function PlotLegends

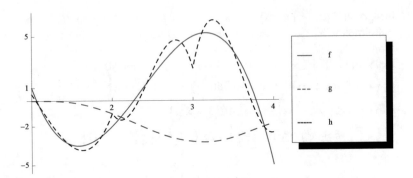

Fig. 5.9 The built-in function PlotLegends with color

We can change the color and style of the previous graph.

```
Plot[{f@x,g@x,h@x},{x,1,4},PlotStyle→{Directive[Red,Thick],
Directive[Thick,Blue,Dashing[{.025}]],
Directive[Thick,Black,Dashing[{.015}]]},
Ticks→{{1,2,3,4},{-5,-2,1,5}},PlotLegend→{f,g,h},
LegendPosition→{1.1,-0.4},ImageSize→450]
```

Above we present the built-in function PlotLegends with color Fig. 5.9.
Define the function

$$f(x) = \begin{cases} \sqrt{x}, & \text{if } x \geq 0, \\ \sqrt{-x}, & \text{if } x \leq 0. \end{cases}$$

We plot its graph for $-1 \leq x \leq 1$.

```
Clear[f]
f[x_?Positive]:=Sqrt@x
f[x_?Negative]:=Sqrt@-x
{Plot[f@x,{x,-1,1},ImageSize→200,Ticks→{{-1,1},{.5,1}}],
Plot[Sqrt@Abs@x,{x,-1,1},ImageSize→200,
Ticks→{{-1,1},{.5,1}}]}
```

The built-in function Sqrt is presented in Fig. 5.10 below.
We can color such a picture. One way of coloring is as follows:

```
Clear[f]
f[x_?Positive]:=Sqrt@x
f[x_?Negative]:=Sqrt@-x
Plot[f@x,{x,-1,1},ImageSize→200,PlotStyle→Thick,
ColorFunction→{Hue@#&},Ticks→{{-1,1},{.5,1}}]
```

Presented below is the built-in function Sqrt with color Fig. 5.11.

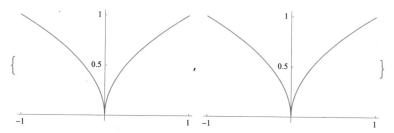

$\left\{ , \right\}$

Fig. 5.10 The built-in function Sqrt

Fig. 5.11 The built-in
function Sqrt with color

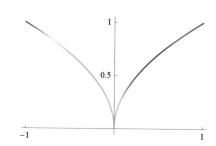

Fig. 5.12 The built-in
function Plot acting on two
functions

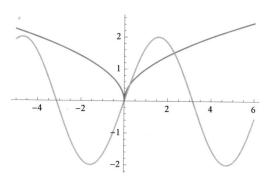

We have seen the usage of the built-in function Plot for several functions. The graphs of different functions are colored with different colors. Here is an example, $f(x)$ defined above and $g(x) = 2\sin x$ on the same interval $[-5, 6]$.

```
Plot[{f@x,2Sin@x},{x,-5,6},ImageSize→250]
```

The built-in function Plot acting on two functions is presented in Fig. 5.12 above. Two other possible ways of coloring are presented below:

```
Plot[{f@x,2Sin@x},{x,-5,6},PlotRange→All,ImageSize→250,#1]&/@
{PlotStyle→{Black,Red},{ColorFunction→Function[{x,y},Hue[y]]}}
```

Two functions colored in different ways are presented in Fig. 5.13.

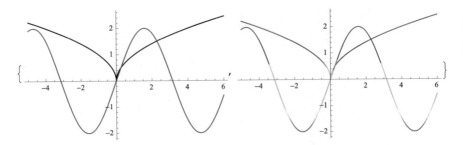

Fig. 5.13 Different colored functions

5.4   Plane and Space Figures

5.4.1   Plane Figures

Mathematica is an excellent tool for drawing figures, simple or very complicated ones.

We note that *Mathematica* considers two rectangular axes such that the desired graph lies in a visible area. The scale of the two axes need not coincide. Maybe we want to have a smaller graph colored in red and dashed. Therefore, we can choose from some options.

```
{Plot[2Sin@t,{t,0,4π},PlotStyle→{Thickness→0.008,Red,Dashed},
AxesOrigin→{0,0},Ticks→ {{0, π, 2π, 3π, 4π},Automatic},
ImageSize→200],
Plot[2Sin@t,{t,0,4 π},PlotStyle→{Thickness→0.008,Blue},
AxesOrigin→ {π,0},AxesLabel→{t,x},
AxesStyle→
{Directive[Thickness→0.01,Green,12],Directive[Purple,12]},
Ticks→ {{0, π, 2π, 3π, 4π},{-2,-1,1,2}},ImageSize→200],
Plot[2Sin@t,{t,0,4π},PlotStyle→{Thickness→0.008,Blue},
AxesOrigin→ {2π,0},AxesLabel→{t,Sin@t},
AxesStyle→{Directive[Thickness→0.01,Green,12],
Directive[Purple,12]},Ticks→ {{0, π, 2π, 3π, 4π},{-2,-1,1,2}},
AspectRatio→Automatic,ImageSize→200]}
```

One way of presentation of functions is introduced in Fig. 5.14 below.

The length of a dash and empty space can be modified. The graph can be exhibited in dotted form.

```
Plot[2Sin@x,{x,-5,7.28},ImageSize→200,
Ticks→{{-5,-2,2,4,7},{-2,-1,1,2}},
PlotStyle→{Thickness→0.008,Black,#1}]&/@
{Dashing[Table[n/100,{n,1,5,0.4}]],Dotted}
```

One way of presentation of functions is introduced in Fig. 5.15 below.

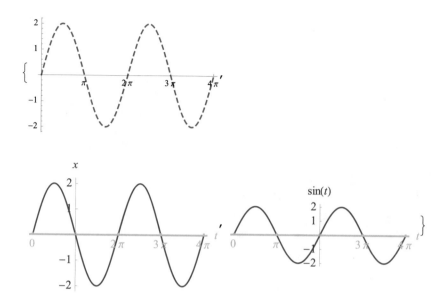

Fig. 5.14 Different presentations of functions 1

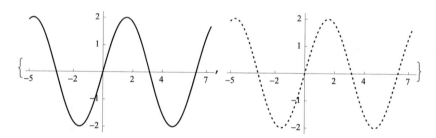

Fig. 5.15 Different presentations of functions 2

We consecutively display and label the graphs of the three functions. At the third plot, the label uses the same font and style that is used in graphs and is not included in the plot.

```
{Plot[3 Cos@x,{x,-5,7.28},
PlotStyle→{Thickness→0.008,Darker[Green],Dashing[0.03]},
PlotLabel→"3cos(x)",ImageSize→180,PlotRange→All],
Plot[2Sin@x,{x,-5,7.28},PlotStyle→{Thickness→0.008,Blue,Dotted},
ImageSize→180,PlotLabel→Style["sin(x)",FontSize→18,
Background→Green],Background→Yellow,PlotRange→All]
Labeled[Plot[Tan@x,{x,-5,7.28},ImageSize→180],
Style["Tan(x)","Graphics"],Background→LightGreen]}
```

Different colors for the pictures are introduced in Fig. 5.16.

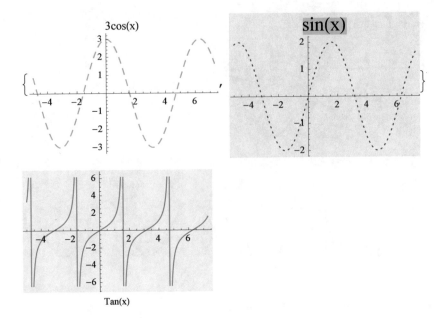

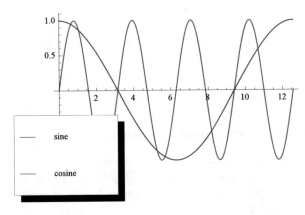

Fig. 5.16 Different colors for the pictures

Fig. 5.17 Different functions
on the same Cartesian system

Suppose we want to put several graphs of functions in the same Cartesian system of coordinates. We can do it in the following way.

```
Plot[{Sin[2x],Cos[x/2]},{x,0,4π},
PlotStyle→{Directive[Thickness→0.007,Red],
Directive[Thickness→0.007,Blue]},PlotLegend→{"sine","cosine"},
ImageSize→300]
```

A way of presentation of functions is introduced in Fig. 5.17 above.
We can also insert the legend exactly where we want.

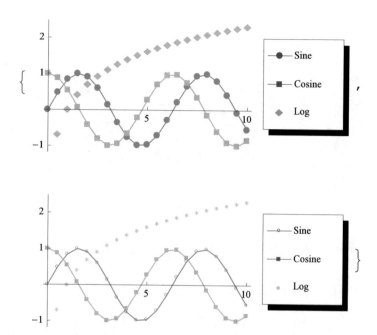

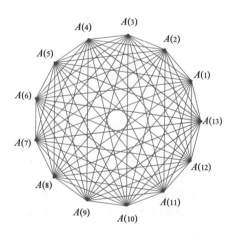

Fig. 5.18 A way of legend insertion

Fig. 5.19 A complete graph
with 13 vertices

```
ListPlot[Table[{x,f@x},{f,Sin,Cos,Log},{x,0,10,0.5}],
PlotLegend→{"Sine","Cosine","Log"},LegendPosition→{1.1,-0.4},
Joined→{True,True,False},Ticks→{{5,10},{-1,1,2}},ImageSize→300,
PlotMarkers→ #1]&/@{Automatic,{"∘","■","*"}}
```

Next, a way of legend insertion is introduced in Fig. 5.18.
A complete graph with 13 nodes is presented in Fig. 5.19. Its vertices are labeled.

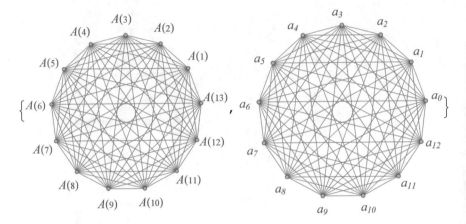

Fig. 5.20 Two complete graphs with 13 vertices

```
Tuples[Table[{Cos@t,Sin@t},{t,0,2π,2π/13}],2];   (* It generates all
vertices of the graph *)
Graphics[{{Blue,Line@%},{Text[Style[Row[{Style[" A(",Italic,12],#1,
Style[" )",Italic,12]}]],{Cos[#1 2π/13],Sin[#1 2π/13]}}&/@Range@13},
ImageSize→200}]
```

The above complete graph can be introduced with a single command Fig. 5.20.

```
Clear[a]
p=CompleteGraph@13;
Show[p,Graphics[{Text[Style[ Row[{Style[" A(",Italic,12],#1,
Style[" )",Italic,12]}]],Red],1.2{Cos[#1 2π /13+π/20],
Sin[#1 2π /13+π/20]}}&/@Range@13],ImageSize→200],
Show[p,Graphics[{Text[Style[a#1−1,Italic,12,Red],
1.15{Cos[(#1-1)2π/13+π/20],Sin[(#1-1) 2π/13+π/20]}}&/@
Range@13],ImageSize→200]
```

See Fig. 5.20.

An old-fashioned hour plate can be designed as follows. Moreover, we can show the exact time at the instant of running the code Fig. 5.21.

```
{hour,minute,second}=
ToExpression[StringSplit[TextString[TimeObject[]],":"]];
minute=minute+second/60;hour=hour+minute/60;hour=Mod[hour,12];
Show[Graphics[{Table[Text[Style["-",Plain,FontWeight→"Bold",Blue,8],
{Cos@θ,Sin@θ}1.01,{0,0},{Cos@θ,Sin@θ}],
{θ,0,2π-2π/60,2π/60}],{Black,PointSize[0.03],Point[{0,0}],
```

Fig. 5.21 An hour plate

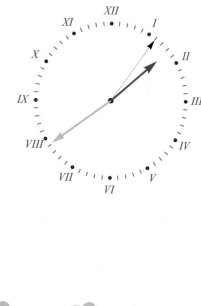

Fig. 5.22 Discs covering a circle and the sine function

```
PointSize[0.02],Point[{Cos[#1π/6],Sin[#1π/6]}],
{Arrowheads[0.05],Thin,Black,Arrow[{{0,0},
.97{Cos[π/2 − π second/30],Sin[π/2 − π second/30]}}]},
{Arrowheads[0.07],Thick,Green,Arrow[{{0,0},
.95{Cos[π/2 − π minute/30],Sin[π/2 − π minute/30]}}]},
{Arrowheads[0.078],Thick,Blue,Arrow[{{0,0},
.8{Cos[π/2 − π hour/6],Sin[π/2 − π hour/6]}}]},
Text[Style[RomanNumeral[#1],Italic,12,Blue],1.16{Cos[(-#1+3)π/6],
Sin[(-#1+3)π/6]}]}]&/@Range@12,ImageSize→200]
```

The color of a figure can vary, thanks to many options. One of them is the graphics directive Hue. Two simple examples follow in Fig. 5.22.

```
Graphics[Table[{Hue[k/7],Disk[{Cos[k 2π/7],Sin[k 2π/7]},1/2]},{k,0,6}],
PlotRange→{{-2,2},{-2,2}},ImageSize→200]
```

We change the previous code so that the discs are laid on the sine function.

```
Graphics[Table[{Hue[k/7],Disk[{k 2π/7,Sin[k 2π/7]},1/2]},{k,0,14}],
PlotRange→{{-1/2,14},{-2,2}},ImageSize→250]
```

See Fig. 5.22.

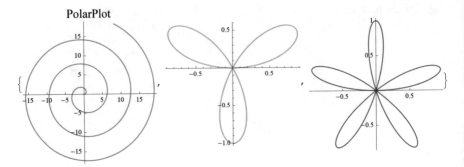

Fig. 5.23 Different spirals

We now present some spirals. In the first example, the polar radius and the angle
are equal. In the second example, the polar radius equals sin(3*t*), whereas in the
third is sin(5*t*). The next two examples show other polar closed curves Fig. 5.23.

```
{PolarPlot[t,{t,0,20},PlotLabel→Style[PolarPlot,16],
Ticks→{{-15,-10,-5,5,10,15},{-15,-10,-5,5,10,15}}],
p3=PolarPlot[Sin[3t],{t,0,π},PlotStyle→{Thickness→0.01,
Darker@Green}],
PolarPlot[Sin[5t],{t,0,π},PlotStyle→{Thickness→0.01,Red},
Ticks→{{-.5,.5},{-.5,.5,1}}]]}

p1=PolarPlot[{1,1/2,1+1/10 Sin[10t]},{t,0,2π},
PlotStyle→{Directive[Thick,Darker@Green],Blue,
Directive[Dashed,Thick,Red]}];
p2=PolarPlot[{1,1-1/5,1+1/5,1+1/5 Sin[10t]},{t,0,2π},
PlotStyle→{Directive[Thick,Darker@Blue],
Directive[Thick,Darker[Black]],Directive[Thick,Darker@Green],
Directive[Dashed,Thick,Red]},Axes→False];
{p1,p2}
Show[p3,p2,PlotRange→All,Axes→ {True,False},ImageSize→200]
```

Some graphs with PolarGraph are shown in Fig. 5.24.
Now we show the simple *cross*, [19, pp. 130–131].

```
With[{a=6,b=4},x[t_]:=a Sec@t; y[t_]:=b Csc@t;
ParametricPlot[{{x@t,y@t},{x[t+π/4],y[t+π/4]}},{t,-π,π},
PlotStyle→Blue,Axes→False,ImageSize→180]
```

We suggest the next code for the *Maltese cross*.

```
PolarPlot[{2/√Sin[4t − π/2],8/√Sin[4t + π/2]},{t,0,2π},Axes→False,
PlotStyle→{Thick,Blue},ImageSize→180]
```

The two crosses are presented in Fig. 5.25.

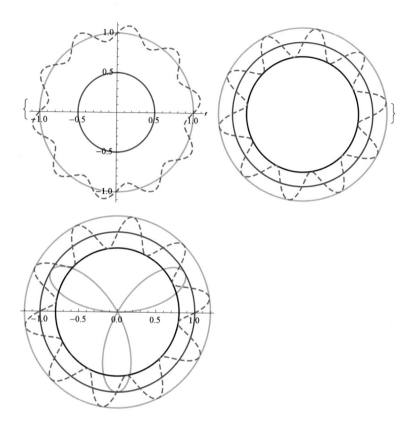

Fig. 5.24 Graphs with polar plot

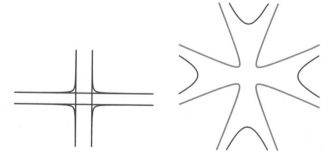

Fig. 5.25 The cross and the Maltese cross

We present the next code for the *Star of David* Fig. 5.26.

```
p=Tuples[Table[{Cos@t,Sin@t},{t,π/6,13π/6,π/3}],2];   (* This is
the complete graph with 6 vertices *)
p=Select[p,EuclideanDistance[#[[1]],#[[2]]]==√3&];   (* From
the complete graph, we select the edges of a prescribed length *)
Show[Graphics[{Thick,Blue,Line[p]}],ImageSize→180]
```

Fig. 5.26 Star of David

Fig. 5.27 Roses

The polar equation of a rhodonea (rose), [19, p. 175], is $r = a\cos(mt)$. We admit that $m \in \mathbb{Z}$. In such a case, there are m petals if m is odd and $2m$ petals if m is even Fig. 5.27.

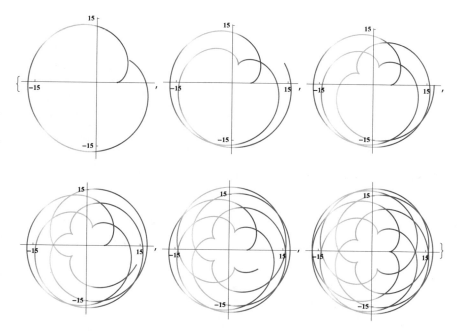

Fig. 5.28 Planar curves by parametric equations, 1

With[{a=4},PolarPlot[a Cos[# t],{t,0,2π},Ticks→{{-a,a},{-a,a}},
TicksStyle→Directive[Red,Bold]]&/@{2,3,4,5}]

We now present some planar curves defined by parametric equations. Giving different domains to the independent variable, we present how the final graphs appear Fig. 5.28.

xx[t_]:=11Cos@t-6Cos[11t/6];
yy[t_]:=11Sin@t-6Sin[11t/6];
ParametricPlot[{xx@t,yy@t},{t,0,#},ColorFunction→{Hue@#&},
Ticks→{{-15,15},{-15,15}},TicksStyle→Directive[Black,Bold]]&/@
{2π,4π,6π,8π,10π,12π}

Other graphs are introduced in Fig. 5.29.

xx[t_]:=11Cos@t-6Cos[√2t/6];
yy[t_]:=11Sin@t-6Sin[√2t/6];
ParametricPlot[{xx@t,yy@t},{t,0,#},ColorFunction→{Hue@#&},
Ticks→{{-15,15},{-15,15}},TicksStyle→Directive[Black,Bold]]&/@
{2π,12π,60π}

The *bean* is a quartic curve given by the implicit equation

$$x^4 + x^2y^2 + y^4 = x\left(x^2 + y^2\right).$$

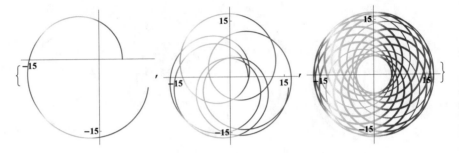

Fig. 5.29 Planar curves by parametric equations, 2

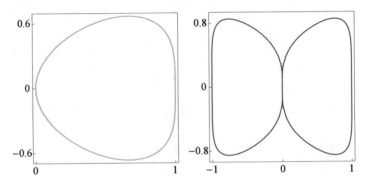

Fig. 5.30 A bean and a simple butterfly

It has horizontal tangents at $(2/3, \pm 2/3)$ and vertical tangents at $(0,0)$ and $(1,0)$. Its graph looks as the left-hand graph in Fig. 5.30.

```
ContourPlot[x⁴+x²y²+y⁴==x(x²+y²),{x,0,1},{y,-2/3,2/3},
ContourStyle→{Darker@Green},FrameTicks→{{{-0.6,0,0.6},None},
{{0,1},None}},ImageSize→180]
```

A simple *butterfly* follows. Its equation is given implicitly, the right-hand graph in Fig. 5.30.

```
ContourPlot[y⁶ == x² − x⁶,{x,-1,1},{y,-.9,.9},ContourStyle→{Blue},
FrameTicks→{{{-0.8,0,0.8},None},{{-1,0,1},None}},
ImageSize→180]
```

A more realistic butterfly curve is a transcendental plane curve discovered by Temple H. Fay in [23]. This curve is given by the following parametric equations Fig. 5.31.

```
xx[t_]:=Sin@t(e^Cos@t-2Cos[4t]-Sin[t/12]⁵);
yy[t_]:=Cos@t(e^Cos@t-2Cos[4t]-Sin[t/12]⁵);
ParametricPlot[{xx@t,yy@t},{t,0,#1},ColorFunction→{Hue@#&},
Ticks→{{-3,0,3},{-2,3}}]&/@{2Pi,4Pi,6Pi,8Pi,12Pi,24Pi}
```

ParametricPlot with two parameters gives a region Fig. 5.32.

Fig. 5.31 Butterflies

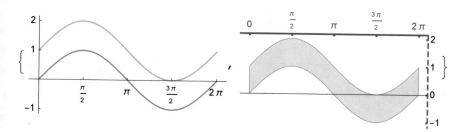

Fig. 5.32 ParametricPlot with two parameters giving a region

```
Apply[{ParametricPlot[{{t,Sin@t},{t,Sin@t+1}},{t,0,2π},
Ticks→ #1,#2],
ParametricPlot[{t,r Sin@t+(1-r)(Sin@t+1)},{t,0,2π},{r,0,1},
Frame→{{False,True},{False,True}},
FrameStyle→{{Thick,Directive[Thick,Dashed]},{Red,Blue}},
FrameTicks→ #1,#2]}&,{{{0,Pi/2,Pi,3 Pi/2,2 Pi},{-1,0,1,2}},
ImageSize→200}]
```

By default the functions in the built-in function Plot are not evaluated until they have numeric values for the variables Fig. 5.33.

```
{Plot[Table[2Sin[x+t],{t,0,4}],{x,0,2π},Ticks→{{π,2π},{-2,2}}],
Plot[Evaluate[Table[2Sin[x+t],{t,0,4}]],{x,0,2π},Ticks→{{π,2π},
{-2,2}}]}
```

Evaluating the argument symbolically can sometimes be faster, as the following example shows Fig. 5.34.

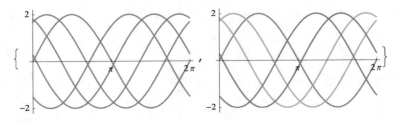

Fig. 5.33 The use of built-in function Evaluate, 1

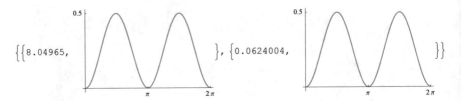

Fig. 5.34 The use of built-in function Evaluate, 2

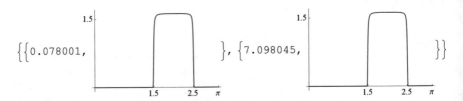

Fig. 5.35 The use of built-in function Evaluate, 3

```
Clear[a,b]
{Timing[Plot[Integrate[Sin@x Cos@x,{x,0,a}],{a,0,2π},
Ticks→{{π,2π},{.5}}]],
Timing[Plot[Evaluate[Integrate[Sin@x Cos@x,{x,0,b}]],{b,0,2π},
Ticks→{{π,2π},{.5}}]]}
```

It is clear that the first graph needs more than 8 s whereas the second one needs about 129 times less.

Sometimes numeric evaluation is faster, as the following example shows Fig. 5.35.

```
{Timing[Plot[Nest[Sin[#]*#&,a,20],{a,0,π},PlotStyle→Blue,
Ticks→{{1.5,2.5,π},{1.5}}]],
Timing[Plot[Evaluate[Nest[Sin[#]*#&,a,20]],{a,0,π},PlotStyle→Blue,
Ticks→{{1.5,2.5,π},{1.5}}]]}
```

We show how the built-in function Nest is used to study the uniform convergence of a sequence of functions Fig. 5.36.

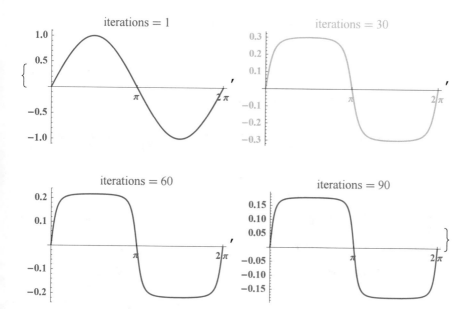

Fig. 5.36 The use of built-in function Nest

```
Plot[Nest[Sin@#&,a,#(10)],{a,0,2π},
  PlotLabel→ToExpression["iterations"]==TraditionalForm[#(10)],
  LabelStyle→Hue[#/10],
  PlotStyle→{Thickness→0.01,Hue[#/10]},
  Ticks→{{π,2π},Automatic},
  TicksStyle→Directive[Bold]]&/@{1/10,3,6,9}
```

We know the behavior of the function $f(x, n) = \sin(\sin(\ldots(\sin x)))$, n times, $x \in [0, \pi]$ Fig. 5.37.

```
Table[
  Plot[Nest[Sin@#&,a,n],{a,0,π},
  PlotRange→All,
  Ticks→{{{π/2,"π/2"},π},Automatic},
  PlotLabel→ToExpression["iterations"]==TraditionalForm[n],
  PlotStyle→{Thickness→0.01,Hue[n/1000]}],
  {n,1,1000,200}]
```

Taking into account that $f(x) = -f(-x),$, we note that $\lim_{n\to\infty} f(x, n) = 0$, uniformly in $x \in \mathbb{R}$. A rigorous proof can be given, too.

At the same time, we have the stronger result that $\lim_{n\to\infty} \sqrt{n/3} f(x, n) = \operatorname{sgn}(x)$, for $x \in]-\pi, \pi[$, Fig. 5.38. A proof is given in the Proposition 5.1.

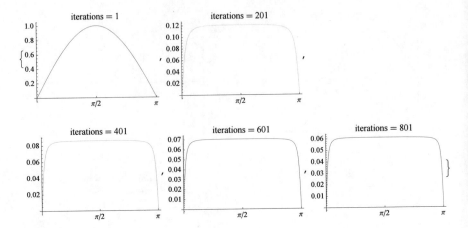

Fig. 5.37 Iteration of the function $f(x, n)$

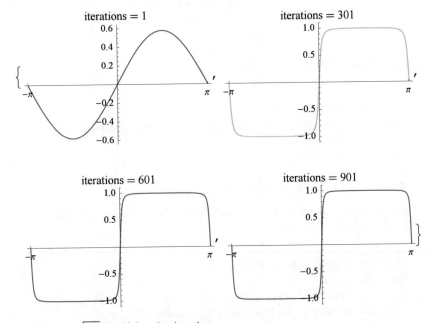

Fig. 5.38 The $\sqrt{n/3} f(x, n))$ function iteration

```
Table[Plot[Sqrt[n/3]×Nest[Sin,x,n],{x,-π,π},
PlotLabel→ToExpression["iterations "]==TraditionalForm[n],
PlotStyle→{Hue[n/1000],Thickness→0.008},
Ticks→{{-π,π},Automatic},PlotRange→All],
{n,1,1000,300}]
```

We can handle the last command dynamically by the next code. Changing by a click the value of n, the color of the graph changes Fig. 5.39.

Fig. 5.39 Dynamical
changing of Fig. 5.38

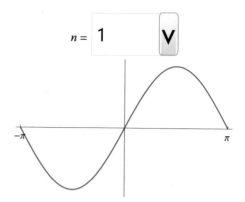

```
Clear[n]
Dynamic[Plot[Sqrt[n/3]×Nest[Sin,x,n],{x,-π,π},ImageSize→200,
Ticks→{{-π,π},{-1,1}},PlotStyle→{Hue[n/1000],Thickness→0.008},
PlotLabel→Row[
{Style[n,Italic]," = ",PopupMenu[Dynamic[n],{1,301,601,901,1201}]}]]]
```

Proposition 5.1 (de Bruijn formula) *Denote as above* $f(x, n) = \sin(\sin(\ldots(\sin x)))$, *n times, $n \in \mathbb{N}^*$, $x \in] - \pi, \pi[$. Then,*

$$\lim_{n \to \infty} \sqrt{n/3} f(x, n) = \operatorname{sgn} x.$$

Proof Obviously $f(x, n + 1) = \sin f(x, n)$, $n \in \mathbb{N}$ and $x \in \mathbb{R}$. Choose $x \in]0, \pi[$. Then $f(x, 1) \in]0, 1] \subset]0, \pi/2]$, $0 \le f(n + 1, x) < f(n, x)$, and $f(n, x) \longrightarrow 0$ uniformly in x. We have

$$\frac{1}{\sin^2 t} - \frac{1}{t^2} = \frac{t - \sin t}{t^3} \times \frac{t}{\sin t} \times \left(\frac{t}{\sin t} + 1 \right) \longrightarrow 1/3 \ \text{as} \ t \downarrow 0.$$

Substitute $t:=f(n, x)$ to get

$$\frac{1}{f^2(n + 1, x)} - \frac{1}{f^2(n, x)} \longrightarrow 1/3 \ \text{as} \ n \longrightarrow \infty.$$

According to the Corollary 1.5 at page 87 in [53], we further have that

$$\frac{1}{n} \sum_{i=1}^{n-1} \left(\frac{1}{f^2(i + 1, x)} - \frac{1}{f^2(i, x)} \right)$$

$$= \frac{1}{n} \left(\frac{1}{f^2(n, x)} - \frac{1}{f^2(1, x)} \right) \longrightarrow 1/3 \ \text{as} \ n \longrightarrow \infty$$

and

$$nf^2(n, x) \longrightarrow 3 \quad \text{as} \quad n \longrightarrow \infty. \qquad \qquad \Box$$

Example The built-in function NestList is a powerful tool in studying sequences defined by recurrences. Below, we introduce two examples.

Suppose we want to study the following sequences defined by recurrences

$$x_1 = \sqrt{2}, \quad x_2 = \sqrt{2 + \sqrt{2}}, \ldots, x_{n+1} = \sqrt{2 + \sqrt{x_n}}, \quad \text{and}$$

$$x_1 = \sqrt{2}, \ldots, x_{n+1} = \left(\sqrt{2}\right)^{\sqrt{x_n}}, \quad n \in \mathbb{N}^*.$$

```
Clear[a,n]
Assumptions→n∈Integers&&n>0;
With[{n=10,a=2.0},
{NestList[√a + #&, √a, n],    (* Exhibit the first ten terms of the first
sequence *)
NestList[(√a)# &, √a, 3n]    (* Exhibit the first thirty terms of the
second sequence *)}]
{{1.41421,1.84776,1.96157,1.99037,1.99759,1.9994,1.99985,1.99996,
1.99999,2.,2.},
{1.41421,1.63253,1.76084,1.84091,1.89271,1.927,1.95003,1.96566,1.97634,
1.98367,1.98871,1.99219,1.99459,1.99626,1.99741,1.9982,1.99876,
1.99914,1.9994,1.99959,1.99971,1.9998,1.99986,1.9999,1.99993,1.99995,
1.99997,1.99998,1.99998,1.99999,1.99999}}
```

We note that the above evaluations suggest that both sequences tend toward 2. The limits are found rigorously in [53, p. 140] and [54, p. 28].   △

Let us color some planar curves with different colors Figs. 5.40 and 5.41.

```
Plot[#1,{t,0,15},PlotStyle→AbsoluteThickness@4,
ColorFunction→"Rainbow",
ImageSize→200]&/@{Sin@t,Sin@t/t,t Sin@t}

{Plot[Sin@t,{t,0,10},PlotStyle→AbsoluteThickness[5],
ColorFunction→"BlueGreenYellow",#1],
Plot[Sin@t/t,{t,0,15},PlotStyle→AbsoluteThickness[4],
ColorFunction→Function[{x,y},Hue[x]],#1],
Plot[t Sin@t,{t,0,10},PlotStyle→AbsoluteThickness[2],
ColorFunction→Function[{x,y},ColorData["NeonColors"][y]],#1]}
&/@{ImageSize→200}}
```

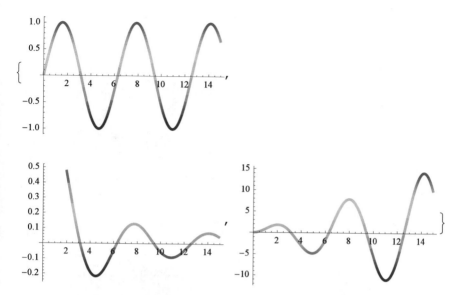

Fig. 5.40 Planar curves with rainbow colors, 1

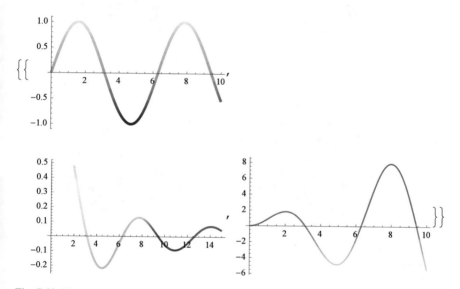

Fig. 5.41 Planar curves with several colors, 2

5.4.2   *Functions and Graphs in 3D*

A surface of revolution can be obtained as the following examples show. Then, we get a surface looking like a UFO Fig. 5.42.

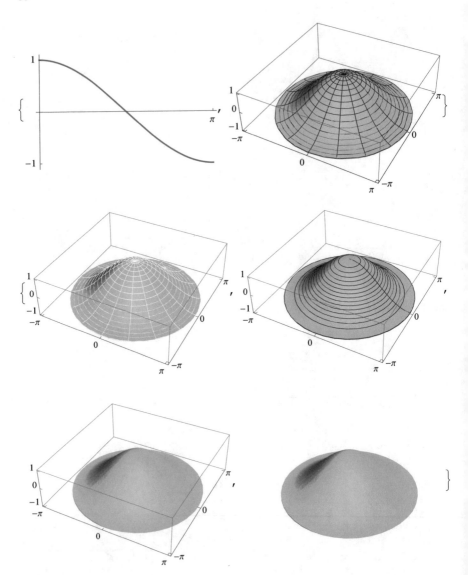

Fig. 5.42 Getting a UFO

meshstyle={MeshStyle→GrayLevel[1],MeshFunctions→{#3&},
Mesh→None,{Mesh→None,Boxed→False,Axes→False}};
{Plot[Cos@t,{t,0,π},PlotStyle→{Thickness→0.01},
Ticks→{{π},{-1,1}}],
RevolutionPlot3D[Cos@t,{t,0,π},PlotStyle→{Green,Opacity→0.4,

Fig. 5.43 A broken line in
the 3D space

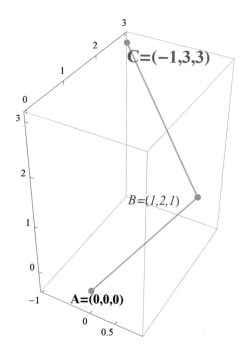

```
Thickness[.02]},ImageSize→200,
Ticks→{{-π,0,π},{-π,0,π},{-1,0,1}}]}
RevolutionPlot3D[Cos@t,{t,0,π},PlotPoints→40,
PlotStyle→{Green,Opacity→0.4,Thickness[.02]},ImageSize→200,
Ticks→{{-π,0,π},{-π,0,π},{-1,0,1}}#]&/@meshstyle
```

A 3D curve very often has several representations. The first example shows a
broken line connecting 3 points in the 3D space Fig. 5.43.

```
Graphics3D[{{Thickness→0.008,Darker[Green],
Line[{{0,0,0},{1,2,1},{-1,3,3}}]},
Text[Style[" A=(0,0,0)",Bold,12],{0,0.2,-0.4}],
Text[Style[" B=(1,2,1)",Italic,12,Darker[Blue]],{0.5,1.3,1.2}],
Text[Style[" C=(-1,3,3)",Bold,16,Red],{0,2.7,3.1}],
{PointSize[0.03],Orange,Point[{{0,0,0},{1,2,1},{-1,3,3}}]}}},
Axes→True, PlotRange→All,ImageSize→200,
Ticks→{{-1,0,0.5},{0,1,2,3},{0,1,2,3}}]
```

We now introduce a cube, its vertices, and some facts about these elements.

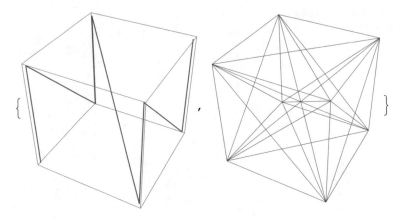

Fig. 5.44 A broken line in the unit cube

```
p=Tuples[{0,1},3]   (* The vertices of the unit cube *)
MatrixForm/@%
{p[[2]],   (* The second element of the list p *)
p[[2,3]]}   (* The third element of the second element in the list p *)
q=Tuples[p,2];   (* This list is large, so we do not display it *)
q[[3]]
q[[3,2]]
```

$\{\{0, 0, 0\}, \{0, 0, 1\}, \{0, 1, 0\}, \{0, 1, 1\}, \{1, 0, 0\}, \{1, 0, 1\}, \{1, 1, 0\}, \{1, 1, 1\}\}$

$$\left\{\begin{pmatrix} 0 \\ 0 \\ 0 \end{pmatrix}, \begin{pmatrix} 0 \\ 0 \\ 1 \end{pmatrix}, \begin{pmatrix} 0 \\ 1 \\ 0 \end{pmatrix}, \begin{pmatrix} 0 \\ 1 \\ 1 \end{pmatrix}, \begin{pmatrix} 1 \\ 0 \\ 0 \end{pmatrix}, \begin{pmatrix} 1 \\ 0 \\ 1 \end{pmatrix}, \begin{pmatrix} 1 \\ 1 \\ 0 \end{pmatrix}, \begin{pmatrix} 1 \\ 1 \\ 1 \end{pmatrix}\right\}$$

$\{\{0, 0, 1\}, 1\}$
$\{\{0, 0, 0\}, \{0, 1, 0\}\}$
$\{0, 1, 0\}$

We now show some facts on the unit cube Fig. 5.44.

```
{Graphics3D[{Red,Thickness→0.008,Line@p}],
Graphics3D[{Blue,Line@q},Boxed→False]}   (* In the first cube, the vertices
are connected following the order of the list p; in the second cube, each
vertex is connected to the other ones *)
```

We plot some functions using different coordinate systems and different label options.

```
{Show[ParametricPlot3D[{(2+Cos@v)Cos@u,(2+Cos@v)Sin@u,
Sin@v},{u,0,2π},{v,π,2π},Mesh→None,
PlotLabel→Style[ParametricPlot3D,Italic,Green,16],
Ticks→{{-3,0,3},{-3,0,3},{-1,0,1}}],
ParametricPlot3D[{Cos@u Sin@v,Sin@u Sin@v,
u v((π/2)²-v²)Cos@v},{u,0,2π},{v,0,π/2},Mesh→None],
ImageSize→200,ViewPoint→{2.5,2.25,2.5},PlotRange→All],
```

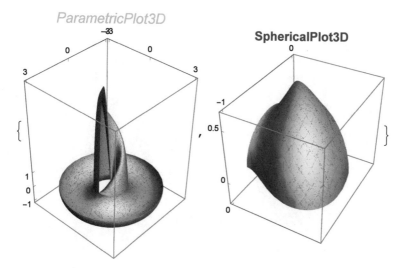

Fig. 5.45 Parametric and spherical plot

```
SphericalPlot3D[Sin@θ+Sin[5φ]/5,{θ,0,π/2},{φ,3π/2,11π/6},
Ticks→{{-1,0,1},{-1,0,1},{-.5,0,.5}},Mesh→None,
PlotLabel→Style[SphericalPlot3D,14,Bold,Blue],ImageSize→200]}
```
See Fig. 5.45.

We now introduce two helices. The first is conic and the second is cylindrical. Then, we put them together into the same frame Fig. 5.46.

```
Clear[r,t]
{p1=ParametricPlot3D[{t Cos@t,t Sin@t,t},{t,0,2π},
PlotStyle→{Directive[Thick,Darker[Green]]},ImageSize→200],
p2=ParametricPlot3D[{{Cos@r,Sin@r,0},{Cos@r,Sin@r,r}},{r,0,2π},
PlotStyle→{Directive[Thick,Red]},ImageSize→{150,275},
Axes→True,AxesOrigin→{0,0,0},Ticks→{{-1,1},{-1,0,1},{π,2π}},
TicksStyle→16,AxesStyle→Directive[Thickness→.0175,
Darker[Green]],ViewPoint→{3,0,π/3},Boxed→False],
Show[p1,p2,ImageSize→200]}
```

The *Möbius surface* is given below Fig. 5.47.

```
{pmoebius=ParametricPlot3D[
{Cos[t](3+r Cos[t/2]),Sin[t](3+r Cos[t/2]),r Sin[t/2]},{r,-1,1},{t,0,2π},
Mesh→None,PlotPoints→{75,100},MaxRecursion→ 0,
PlotStyle→{Green,Opacity→0.4,Thickness[.02]},
PlotPoints→50,Ticks→{{-3,0,2,4},{-3,0,3},{-1,0,1}},
ImageSize→250],
Show[pmoebius,Boxed→False,Axes→False,ImageSize→250]}
```

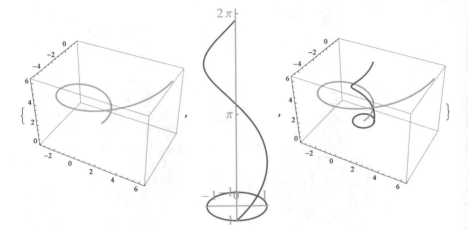

Fig. 5.46 Conic and cylindrical helices

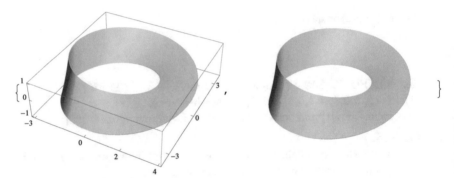

Fig. 5.47 Möbius surface

Now the edge of the Möbius surface follows Fig. 5.48.

```
eqs[r_,t_]:={Cos[t](3+r Cos[t/2]),Sin[t](3+r Cos[t/2]),r Sin[t/2]};
ParametricPlot3D[{eqs[1,t],eqs[-1,t]},{t,0,2π},
Ticks→{{-2,4},{-2,2},{-1,1}},
PlotStyle→{Blue,Dashed,Thickness[.2]},ImageSize→250]
```

We now plot a function in different coordinate systems Fig. 5.49.

```
{Plot3D[Sin[x y],{x,0,3},{y,0,3},Mesh→#,
Ticks→{Automatic,Automatic,{-1,0,1}}],
DensityPlot[Sin[x y],{x,0,3},{y,0,3},Mesh→#]}&/@{All,None}
```

An example of a saddle point on a 3D surface (hyperbolic paraboloid) is presented below Fig. 5.50.

```
Show[ParametricPlot3D[{x,y,(x²-y²)/2},{x,-2,2},{y,-2,2},
Boxed→False,Axes→False,MeshStyle→GrayLevel[1],
PlotStyle→{Lighter[Green],Opacity→0.6},PlotPoints→{100,60}],
```

Fig. 5.48 The edge of the
Möbius surface

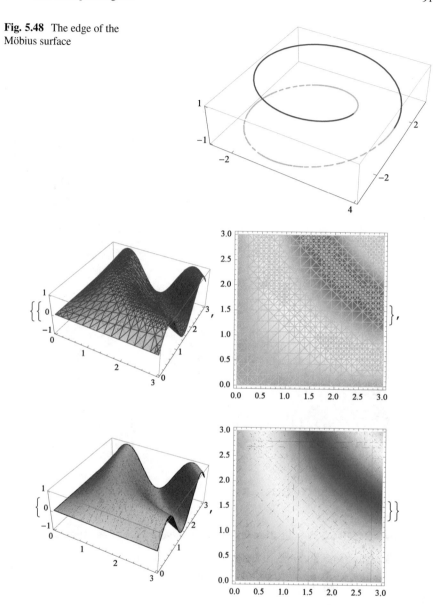

Fig. 5.49 Different coordinate systems

```
ParametricPlot3D[{x,0,x²/2},{x,-2,2},
PlotStyle→{Directive[Thick,Darker[Red]]}],
ParametricPlot3D[{0,y,-y²/2},{y,-2,2},
PlotStyle→{Directive[Thickness→0.006,Darker[Red]]}],
Graphics3D[{PointSize[0.04],Black,Point[{{0,0,0}}]}],
PlotLabel→Style["Saddle point of z = x²/2 − y²/2",FontSize→14,
Background→Yellow],ImageSize→250]
```

Fig. 5.50 Hyperbolic
paraboloid

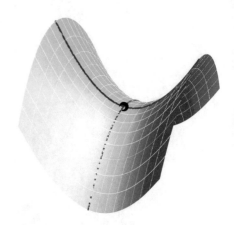

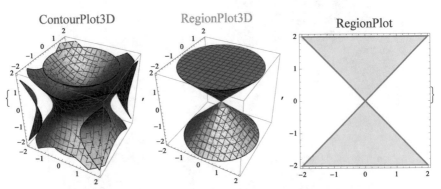

Fig. 5.51 Examples with ContourPlot, RegionPlot3D, and RegionPlot

A few specialized visualization functions are given in the next sequence Fig. 5.51.

```
{ContourPlot3D[x²+y²-z²,{x,-2,2},{y,-2,2},{z,-2,2},
PlotLabel→Style[ContourPlot3D,Red,16]],
RegionPlot3D[x²+y² ≤ z²,{x,-2,2},{y,-2,2},{z,-2,2},
PlotLabel→Style[RegionPlot3D,Orange,16]],
RegionPlot[x² ≤ y²,{x,-2,2},{y,-2,2},
PlotLabel→Style[RegionPlot,Blue,16]]}
```

Chapter 6
Manipulate

6.1 Manipulate

The Manipulate command is very useful, allowing a dynamical approach to results and corresponding figures.

Along this book we will extensively use the Manipulate command concurrent by the Table command.

    Table[n,{n,15}]
    Table[Prime@n,{n,15}]   (* the nth prime number *)
    {1,2,3,4,5,6,7,8,9,10,11,12,13,14,15}
    {2,3,5,7,11,13,17,19,23,29,31,37,41,43,47}

We can now use the Manipulate command as the next example shows Fig. 6.1.

    Manipulate[Prime@n,{n,1,15,1}]

See Fig. 6.1.

In our example in the Table command, the variable n takes discrete values. In this case in the Manipulate command, we require the same variable to be discrete, too.

Let us compare the next two cases. In the first case n is a positive integer, whereas in the second case n is a real positive number within the interval [1, 15].

    {Manipulate[n!,{n,1,11,1}],Manipulate[n!,{n,1,11}]}

See Fig. 6.2.

Another simple example regarding the Manipulate command is the following:

    {Manipulate[#,{n,1,15,1}],Manipulate[#,{n,1,15}]}&/@
    {Plot[Sin[n x],{x,0,2π},ImageSize→150,Ticks→{{π,2π},{-1,1}}]]}

See Fig. 6.3.

The previous example is improved by adding the option

Appearance→"Labeled". This option allows us to see the instant value of the control variable.

© Springer International Publishing AG 2017
M. Mureşan, *Introduction to Mathematica*® *with Applications*,
DOI 10.1007/978-3-319-52003-2_6

Fig. 6.1 Manipulate 1

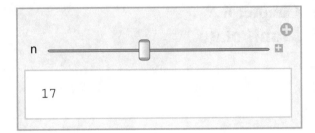

Fig. 6.2 Manipulate 2

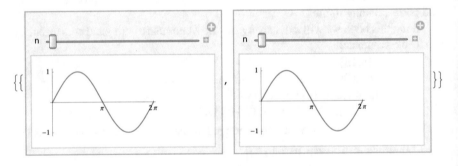

Fig. 6.3 Manipulate 3

Manipulate[Plot[Sin[n x],{x,0,2π},ImageSize→150,
Ticks→{{π,2π},{-1,1}}],{n,1,15,Appearance→"Labeled"}]

See Fig. 6.4.

We can also use several control variables as the following example shows:

Manipulate[Plot[Sin[n x]+Sin[m x],{x,0,2π},PlotRange→2,
ImageSize→200,Ticks→{{π,2π},Automatic}],
{n,1,10,Appearance→"Labeled"},{m,1,15,Appearance→"Labeled"}]

See Fig. 6.5.

For a discrete variable, as we did above, let us consider for example *Newton's binomial formula*.

Manipulate[Expand[$(\alpha + \beta)^n$],{n,1,15,1,Appearance→"Labeled"}]

See Fig. 6.6.

Fig. 6.4 Manipulate 4

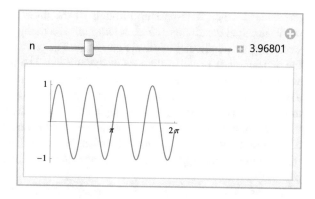

Fig. 6.5 Manipulate 5

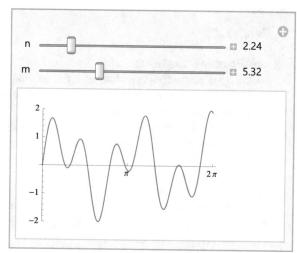

Fig. 6.6 Manipulate 6

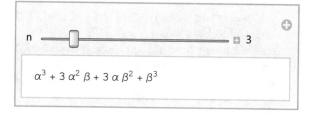

6.1.1   Circumcircle, Incircle, and the Main Points in a Triangle

A *perpendicular bisector* of a side of a triangle is a straight line passing through the midpoint of the side and being perpendicular to it. The three perpendicular bisectors meet in a single point, the triangle's *circumcenter*. This point is the center of the *circumcircle*, the circle passing through all the three vertices. The circumcircle's radius is called the *circumradius*. An *altitude* of a triangle is a straight line through a vertex and perpendicular to the opposite side. The point where the altitude intersects the opposite side is called the *foot* of the altitude. The three altitudes intersect in a single point called the *orthocenter* of the triangle. An *angle bisector* of a triangle is a straight line through a vertex which cuts the corresponding angle into halves. The three angle bisectors intersect in a single point called the *incenter*. This is the center of the triangle's *incircle*. The incircle is the circle which lies inside the triangle and touches all three sides. A *median* of a triangle is a straight line through a vertex and the midpoint of the opposite side. The three medians intersect in a single point, the triangle's *centroid*.

We may assume that vertex *A* of a triangle is free, whereas the vertices *B* and *C* are fixed.

```
Clear[a1,a2]

mainPointsinTriangle[a1_,a2_]:=
Module[{b1=b2=c2=0,c1=6,a,b,c,x,y,ahax,ahay,aha,bhbx,bhby,bhb,
chcx,chcy,chc,op,ox,oy,hx,hy,hp,centroidx,centroidy,centroid,aleft,
aright,bix,biy,bip,bbprime,rRadius},
vertexA={a1,a2};vertexB={b1,b2};vertexC={c1,c2};
a=EuclideanDistance[vertexB,vertexC];
b=EuclideanDistance[vertexC,vertexA];
c=EuclideanDistance[vertexA,vertexB];
If[a≥b+c‖b≥c+a‖c≥a+b,Print[" No triangle"],
centroid={centroidx,centroidy}=(vertexA+vertexB+vertexC)/3;
midAB=(vertexA+vertexB)/2;midBC=(vertexB+vertexC)/2;
midCA=(vertexC+vertexA)/2;
aha={ahax,ahay}={a1,0};
sol2=Solve[
y-b2==-(a1-c1)(x-b1)/(a2-c2)&&y-c2==(a2-c2)(x-c1)/(a1-c1),{x,y}];
bhb={bhbx,bhby}={x,y}/.sol2[[1]];
sol3=Solve[y-c2==-a1(x-c1)/a2&&y==a2 x/a1,{x,y}];
chc={chcx,chcy}={x,y}/.sol3[[1]];
sol4=Solve[x==a1&&{x,y}∈InfiniteLine[{vertexB,bhb}],{x,y}];
hp={hx,hy}={x,y}/.sol4[[1]];
sol5=Solve[
x==(b1+c1)/2&&(y-(c2+a2)/2)==-(a1-c1)(x-(c1+a1)/2)/(a2-
c2),{x,y}];
op={ox,oy}={x,y}/.sol5[[1]]; {aleft,aright}=a{c,b}/(b+c);
bip=vertexC+{a1-c1,a2}a/(a+c);
sol7=Solve[y==a2(x-aleft)/(a1-aleft)&&y==
(c2+a a2/(a+c))x/(c1+a(a1-c1)/(a+c)),{x,y}];
```

```
bisec={bisx,bisy}={x,y}/.sol7[[1]];
sol8=Solve[y==a2 x/a1&&y-c2==(bisy-c2)(x-c1)/(bisx-c1),{x,y}];
cisec={x,y}/.sol8[[1]];rRadius=Norm[op]; rinradius=Divide[a a2,a+b+c];
abλ=vertexA+((vertexA-vertexB).(vertexA-bisec))(vertexB-vertexA)/c²;
bcλ=vertexB+((vertexB-vertexC).(vertexB-bisec))(vertexC-vertexB)/a²;
caλ=vertexC+((vertexC-vertexA).(vertexC-bisec))vertexA-vertexC)/b²;
Show[Graphics[{Text[Style[" A",Italic,12],vertexA+{.01,0.3}],
Text[Style[" B",Italic,12],vertexB-{.2,.15}],
Text[Style[" C",Italic,12],vertexC-{-.2,.15}],
Text[Style[" Hₐ",Italic,12],{ahax,b2-.2}],
Text[Style[" Iₐ",Italic,12],{aleft,b2-.2}],
Text[Style[" Mₐ",Italic,12],{(b1+c1)/2,b2-.2}],
Blue,Line[{{vertexA,vertexB},{vertexB,vertexC},
{vertexC,vertexA},{vertexA,aha},{vertexC,chc},{vertexB,bhb},
{vertexC,midAB},{vertexA,midBC},{vertexB,midCA},
{vertexA,op},{vertexB,op},{vertexC,op}}],
Darker[Green],Line[{{vertexA,{aleft,0}},{vertexB,bip},
{vertexC,cisec},{bisec,abλ},{bisec,bcλ},{bisec,caλ}}],
Black,Line[{{op,midAB},{op,midBC},{op,midCA}}],
Black,PointSize[.015],Point[{vertexA,vertexB,vertexC,bip,aha,bhb,
chc,midAB,midBC,midCA,{aleft,0},cisec,abλ,bcλ,caλ}],
Red,Point[{hp,op,centroid,bisec}],Circle[bisec,rinradius],
Circle[op,rRadius]}],ImageSize→300,PlotRange→All]]]
```

We can now see the important points of a triangle as well as the circumcircle and the incircle change. H_A, I_A, and M_A are the feet of, respectively, the height, bisector, and median of vertex A.

```
Manipulate[
Quiet@mainPointsinTriangle[a1,a2],
{{a1,1,"abscissa"},1,6,0.0001,Appearance→"Labeled"},
{{a2,4,"ordinate"},4,10,0.001,Appearance→"Labeled"},
SaveDefinitions→True,SynchronousUpdating→False
]
```

See Fig. 6.7.

6.1.2 Euler's Nine Points Circle

Euler's nine points circle is an interesting circle defined by nine points related to a triangle. The midpoints of the three sides and the feet of the three altitudes all lie on a single circle, the triangle's nine-point circle. The remaining three points for which it is named so are the midpoints of the segments of altitudes between the vertices and the orthocenter. The radius of the nine-point circle is half of the circumradius.

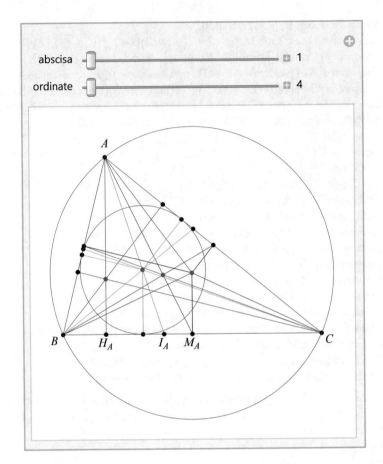

Fig. 6.7 Main points in a triangle

```
Clear[a1,a2]

eulersNinePointsCyrcle[a1_,a2_]:=
Module[{b1=b2=c2=0,c1,a,b,c,x,y,ap,ahax,ahay,bp,bhbx,bhby,cp,
chcx,chcy,op,ox,oy,hx,hy,hp,radius},
vertexA={a1,a2};vertexB={b1,b2};vertexC={c1,c2};
a=EuclideanDistance[vertexB,vertexC];
b=EuclideanDistance[vertexC,vertexA];
c=EuclideanDistance[vertexA,vertexB];
If[a≥b+c||b≥c+a||c≥a+b,Print["No triangle"],
centroid={centroidx,centroidy}=(vertexA+vertexB+vertexC)/3;
midAB=(vertexA+vertexB)/2;midBC=(vertexB+vertexC)/2;
midCA=(vertexC+vertexA)/2;ap={ahax,ahay}={a1,0};
```

```
sol2=Solve[y-b2==-(a1-c1)(x-b1)/(a2-c2)&&y-c2==
(a2-c2)(x-c1)/(a1-c1),{x,y}];bp={bhbx,bhby}={x,y}/.sol2[[1]];
sol3=Solve[y-c2==-a1(x-c1)/a2&&y==a2 x/a1,{x,y}];
cp={chcx,chcy}={x,y}/.sol3[[1]];
sol4=Solve[x==a1&&{x,y}∈InfiniteLine[{vertexB,bp}],{x,y}];
hp={hx,hy}={x,y}/.sol4[[1]];
sol5=Solve[x==(b1+c1)/2&&(y-(c2+a2)/2)==
-(a1-c1)(x-(c1+a1)/2)/(a2-c2),{x,y}];
op={ox,oy}={x,y}/.sol5[[1]];
radius=EuclideanDistance[midBC,(op+hp)/2];
Show[Graphics[{Text[Style[A,Italic,12],vertexA+{.01,.2}],
Text[Style[B,Italic,12],vertexB-{.2,.15}],
Text[Style[C,Italic,12],vertexC-{-.2,.15}],
Text[Style["H_A",Italic,12],{ahax,b2-.2}],
Text[Style["M_A",Italic,12],{(b1+c1)/2,b2-.2}],
Black,PointSize[.015],Point[{vertexA,vertexB,vertexC,hp,op,centroid}],
Red,Point[{ap,bp,cp,midAB,midBC,midCA,(hp+vertexA)/2,
(hp+vertexB)/2,(hp+vertexC)/2}],PointSize[.02],Point[(hp+op)/2],
Blue,Line[{{vertexB,vertexA},{vertexB,vertexC},{vertexC,vertexA},
{vertexA,ap},{vertexC,cp},{vertexB,bp},{vertexC,midAB},{vertexA,
midBC},{vertexB,midCA},{vertexA,op},{vertexB,op},{vertexC,op},
{op,hp},{vertexA,(hp+op)/2},{vertexB,(hp+op)/2},{vertexC,
(hp+op)/2}}],Red,Circle[(op+hp)/2,radius]}],PlotRange→All]]]
```

In the next example, we consider that the vertices B and C are fixed, whereas the vertex A is free, i.e., its coordinates are free. Thus, we can see how the Euler's circle varies according to A.

```
Manipulate[
Quiet@eulersNinePointsCyrcle[a1,a2],
{{a1,1,"abscisa"},1,6,0.0001,Appearance→"Labeled"},
{{a2,4,"ordinate"},3,7,0.001,Appearance→"Labeled"},
SaveDefinitions→True,SynchronousUpdating→False]
```

See Fig. 6.8.

6.1.3 Frenet–Serret Trihedron of a Helix

The *Frenet–Serret trihedron* describes the kinematic properties of a particle moving along a continuously differentiable curve in the three-dimensional Euclidean space $\mathbb{R}^3$ or the geometric properties of the curve itself irrespective of any motion. More specifically, the formulas describe the derivatives of the so-called tangent, normal, and binormal unit vectors in terms of each other. We assume that

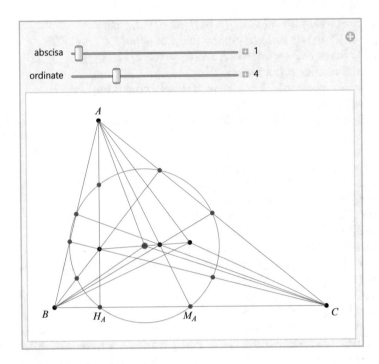

Fig. 6.8 Euler's nine points circle

$x'(t)^2 + y'(t)^2 + z'(t)^2 \neq 0$ for all t, where $r(t) = (x(t), y(t), z(t))$ are the Cartesian coordinates of a point $r(t)$ on the curve. Its arc length is

$$s(t) = \int_{t_0}^t \sqrt{[x'(u)]^2 + [y'(u)]^2 + [z'(u)]^2} \, du = \int_{t_0}^t \left\| r'(u) \right\| du.$$

If $x'(t)^2 + y'(t)^2 + z'(t)^2 \neq 0$ for all t in an interval, the curve is said to be *smooth* on that interval. If the curve is smooth, the arc length function $s(t)$ is increasing, and thus it has an inverse of the form $t = t(s)$. Then the curve r can be reparametrized $r(s)$. In that case s is said to be the *natural parameter* of the curve.

The *unit tangent vector* is defined by

$$\text{tangent}(t) = r'(t) / \left\| r'(t) \right\|$$

and its *curvature* by

$$\kappa = \left\| \frac{d\,\text{tangent}(s)}{ds} \right\| = \frac{\|\text{tangent}'(t)\|}{\|r'(t)\|} = \frac{\|r'(t) \times r''(t)\|}{\|r'(t)\|^3}.$$

We check that the circular helix has a constant curvature.

```
Clear[a,k,r,t]
$Assumptions=a>0&&k>0&&r∈Reals&&t∈Reals;
r[t_]:={a Cos@t,a Sin@t,k t};   (* Circular helix with the usual
representation *)
Norm[Cross[r'@t,r''@t]]/Norm[r'@t]³//Simplify   (* Curvature *)
```
$$\frac{a}{a^2+k^2}$$

We get the same result by the built-in function ArcCurvature.

```
ArcCurvature[r@t,t]//Simplify
```
$$\frac{a}{a^2+k^2}$$

The Frenet–Serret trihedron is discussed in the last chapter where the Viviani's window is exhibited (Sect. 11.1.2). The code presented here is rather elementary.

```
Clear[a,k,s,circle,helix,vt,vs,tangent,tprime,normal,binormal,sfinal,
vt,vs,t]
$Assumptions=a>0&&k>0&&s∈Reals&&t∈Reals;
vt[t_]:={a Cos@t,a Sin@t,k t};   (* Circular helix with the usual
representation *)
```
$$vs[s\_]:=\left\{a\,\cos\left[\tfrac{s}{\sqrt{a^2+k^2}}\right], a\,\sin\left[\tfrac{s}{\sqrt{a^2+k^2}}\right], \tfrac{k\,s}{\sqrt{a^2+k^2}}\right\}   (*\ Circular$$
helix by the arc parameter; this form is useful when we discuss certain
elements of it: tangent, normal, binormal, etc. *)
```
tangent[s_]:=vs'@s   (* Colored in green *)
tprime[s_]:=vs''@s
normal[s_]:=Simplify[tprime@s/Norm[tprime@s]]   (* Colored in red *)
binormal[s_]:=Cross[tangent[s],normal[s]]   (* Colored in blue *)
a=1;k=a/4;   (* Values of the constants *)
sfinal=4π √a² + k²;   (* Final value of the arc parameter *)
```

```
helix=ParametricPlot3D[Evaluate[vs@s],{s,0,sfinal},Axes→None,
Boxed→ False];
circle=
```
$$ParametricPlot3D\left[Evaluate\left[\left\{a\,\cos\left[\tfrac{s}{\sqrt{a^2+k^2}}\right], a\,\sin\left[\tfrac{s}{\sqrt{a^2+k^2}}\right], 0\right\}\right],\right.$$
$$\left.\left\{s, 0, 2\pi\,\sqrt{a^2+k^2}\right\}\right];$$

```
Animate[
Show[helix,circle,Graphics3D[{{PointSize[.02],Black,Point[vs@s]},
{Arrowheads[0.03],Thick,Darker[Green],
Arrow[{vs@s,vs@s+tangent@s}],Red,
Arrow[{vs@s,vs@s+normal@s}],Blue,
Arrow[{vs@s,vs@s+binormal@s}]}}],
ImageSize→180],{s,0,sfinal},AnimationRate→.01,
AnimationRunning→False,SaveDefinitions→True]
```
See Fig. 6.9.

Fig. 6.9 Frenet-Serret
trihedron for a helix

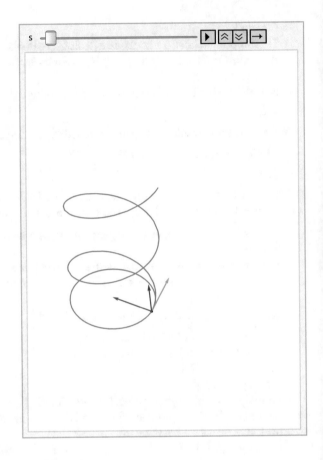

Here we have approached the Frenet–Serret trihedron for the helix by an elementary way. There exists also the built-in function FrenetSerretSystem. This built-in function will be used later for the study of Viviani's window.

6.1.4 Hyperboloid of One Sheet

Now we will introduce the *hyperboloid of one sheet*. This surface models a cooling tower. A hyperboloid cooling tower was patented by the Dutch engineers Frederik van Iterson and Gerard Kuypers in 1918. The general equation of the hyperboloid of one sheet is

$$\frac{x^2}{a^2} + \frac{y^2}{b^2} - \frac{z^2}{c^2} = 1, \quad a > 0,\ b > 0,\ c > 0.$$

Each hyperboloid of one sheet admits two families of rectilinear generators. These generators are of the form

$$x/a - z/c = \lambda(1 - y/b), \quad x/a + z/c = (1/\lambda)(1 + y/b),$$

and

$$x/a - z/c = \mu(1 + y/b), \quad x/a + z/c = (1/\mu)(1 - y/b),$$

where λ and μ are real parameters. Obviously, each generator (straight line) of each family lies entirely on the surface. We find the generators depending on the variable z.

```
Clear[a,b,c,x,y,z,λ,μ]
Solve[x/a-z/c==λ(1-y/b)&&λ(x/a+z/c)==1+y/b,{x,y}]
Solve[x/a-z/c==μ(1+y/b)&&μ(x/a+z/c)==1-y/b,{x,y}]
```

$$\left\{\left\{x \to \frac{a(z+2c\lambda-z\lambda^2)}{c(1+\lambda^2)}, \quad y \to \frac{b(-c+2z\lambda+c\lambda^2)}{c(1+\lambda^2)}\right\}\right\}$$

$$\left\{\left\{x \to \frac{a(z+2c\mu-z\mu^2)}{c(1+\mu^2)}, \quad y \to -\frac{b(-c+2z\mu+c\mu^2)}{c(1+\mu^2)}\right\}\right\}$$

Thus, its rectilinear generators have the equations

$$x = \frac{a\left(z + 2c\lambda - z\lambda^2\right)}{c\left(1 + \lambda^2\right)}, \quad y = \frac{b\left(-c + 2z\lambda + c\lambda^2\right)}{c\left(1 + \lambda^2\right)}, \quad z = z,$$

this is a straight line depending on the parameter λ

$$x = \frac{a\left(z + 2c\mu - z\mu^2\right)}{c\left(1 + \mu^2\right)}, \quad y = -\frac{b\left(-c + 2z\mu + c\mu^2\right)}{c\left(1 + \mu^2\right)}, \quad z = z,$$

this is a straight line depending on the parameter μ.

For an application, we consider below the elliptic hyperboloid of one sheet of equation $z^2 = 4x^2 + 4y^2 - 4$, and its two rectilinear generators, for $z \in [-3, 2]$. Suppose we introduce some particular values for λ and μ. Then their graphs given below are obvious.

```
Clear[λ, μ]

rectilinearGeneratorsHyperboloid[λ_, μ_]:=
Module[{a=b=1,c=2,denominator1=Divide[1,c(1+λ²)],
denominator2=Divide[1,c(1+μ²)],t,u,z},
Show[ParametricPlot3D[{a Sqrt[1+u²]Cos[t],b Sqrt[1+u²]Sin[t],c u},
{t,0,2π},{u,-3/c,2/c},
```

Fig. 6.10 Elliptic
hyperboloid

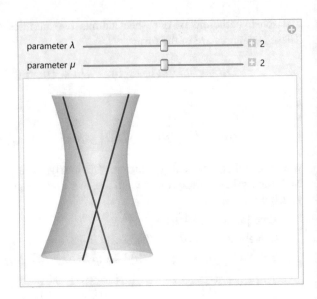

PlotStyle→{Directive[Opacity[0.3],Lighter[Green],PlotPoints→200]},
MaxRecursion→12,Mesh→None,Boxed→False,
ViewPoint→{15,0,1},Axes→False],
ParametricPlot3D[{a(z+2cλ-zλ²)denominator1,
b(-c+2zλ+cλ²)denominator1,z},{z,-3,2},PlotStyle→Black],
ParametricPlot3D[{a(z+2cμ-zμ²)denominator2,
-b(-c+2zμ+cμ²)denominator2,z},{z,-3.1,1.95},PlotStyle→Black],
PlotRange→All,ImageSize→150]]

Manipulate[
Quiet@rectilinearGeneratorsHyperboloid[λ, μ],
{{λ,2," parameter"},-130,130,.1,Appearance→"Labeled"},
{{μ,2," parameter"},-140,140,.1,Appearance→"Labeled"},
SaveDefinitions→True,SynchronousUpdating→False]

See Fig. 6.10.

Chapter 7
Ordinary Differential Equations

Two tutorials issued by Wolfram Research are very helpful along this chapter, namely, [74] and [75].

7.1 Simple Differential Equations

One of the simplest ordinary differential equations is $x'(t) = x(t)$. It is known that if $x : [a, b] \to \mathbb{R}$ and is smooth enough, its general solution has the form $x(t) = ce^t$, where c is a constant. We admit that the constant is a real number. A good source for ordinary differential equations is [31]

```
Clear[t,x]
sol=DSolve[{x'[t]==x[t]},x[t],t]
{{x[t] → e^tC[1]}}
```

We pick out a specific solution using /. (ReplaceAll).

```
y[t_]=x[t]/.sol
Plot[y[t]/.{C[1]→1},{t,-1,1},ImageSize→180,Ticks→{{-1,0,1},{1,E}}]
{e^tC[1]}
```

See Fig. 7.1.

In the next example, we study an initial value problem for a differential equation of the first order.

© Springer International Publishing AG 2017
M. Mureşan, *Introduction to Mathematica® with Applications*,
DOI 10.1007/978-3-319-52003-2_7

Fig. 7.1 A simple ordinary
differential equation

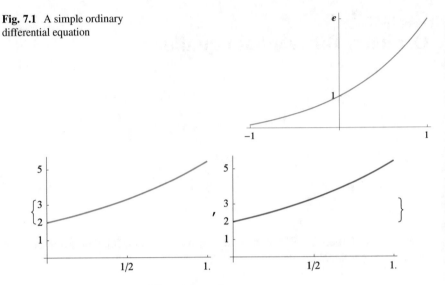

Fig. 7.2   Other simple ordinary differential equations

```
Clear[x,y]
Flatten[
{sol=DSolve[{x'[t]==x[t],x[0]-2==0},x[t],t],eqn={y'[t]==y[t],y[0]-2==0};
sol1=DSolve[eqn,y,t]   (* Here function y is given without argument and
the output form is changed *)},1]
Flatten[{Plot[x[t]/.sol,{t,0,1},#],Plot[y[t]/.sol1,{t,0,1},#,
PlotStyle→Red]}&/@{{AxesOrigin→{0,0},ImageSize→150,
Ticks→{{{0.5,"1/2"},1.},{1,2,3,5}}}}]
{{x[t]→ 2 e^t},{y→Function[{t},2 e^t]}}
```
See Fig. 7.2.

7.2   Systems of Ordinary Linear Homogeneous Differential Equations

We consider the following matrix with real entries $ma = \{\{1, -2, 2\}, \{1, 4, -2\}, \{1, 5, -3\}\}$ and present some solutions.

```
Clear[ma,t,x,y,z,λ]
ma={{1,-2,2},{1,4,-2},{1,5,-3}};
CharacteristicPolynomial[ma,λ];
Eigenvalues[N[ma]]   (* Matrix ma has real and distinct eigenvalues *)
```

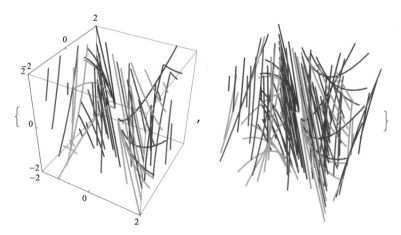

Fig. 7.3 A system of simple ordinary differential equations

```
sol=Flatten[DSolve[{x'[t]==x[t]-2y[t]+2z[t],y'[t]==x[t]+4y[t]-2z[t],
z'[t]==x[t]+5y[t]-3z[t]},{x[t],y[t],z[t]},t]];
x[t_]=sol[[1,2]]; y[t_]=sol[[2,2]]; z[t_]=sol[[3,2]];
functionsarray=Table[{x[t],y[t],z[t]}/.{C[1]→i,C[2]→j,C[3]→k},
{i,-5,5,2},{j,-5,5,2},{k,-2,2,2}];
tograph=Flatten[functionsarray,1];
ParametricPlot3D[Evaluate[tograph],{t,-2,2},
PlotStyle→{Blue,Red,Darker[Green]},Ticks→{{-2,0,2},{-2,0,2},
{-2,0,2}},PlotRange→{{-2,2},{-2,2},{-2,2}}],
ParametricPlot3D[Evaluate[tograph],{t,-2,2},
PlotStyle→{Blue,Red,Darker[Green]},Boxed→False,
Axes→False,PlotRange→{{-3,3},{-3,3},{-3,3}}]
{2.,-1.,1.}
```

See Fig. 7.3.

The next code shows how the same system as an initial value problem looks like. The solution is a curve in the solid space.

```
Clear[x,y,z,t]
A={{1,-2,2},{1,4,-2},{1,5,-3}};
X[t_]={x@t,y@t,z@t};
system=MapThread[#1==#2&,{X'[t],A.X[t]}];
sol=DSolve[Join[system,{x@0==1,y@0==2,z@0==-1}],{x,y,z},t];
ParametricPlot3D[{x[t],y[t],z[t]}/.sol,{t,-2,1},
ColorFunction→Function[{x,y,z},Hue[z]],PlotStyle→{Thick},
PlotRange→All,Ticks→{{0,20},{-20,0,20},{-40,-20,0,20}},
ImageSize→140]
```

See Fig. 7.4.

Fig. 7.4 An initial value
problem

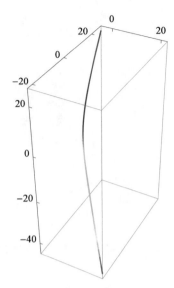

An example of an initial value problem of a system of linear differential with
constant coefficients whose characteristic equation has one real root and two
complex roots is introduced below.

```
Clear[x,y,z]
A={{-1,-2,2},{-2,-1,2},{-3,-2,3}};
Eigenvalues[A];   (* One real and two complex eigenvalues, {1,i,-i} *)
X[t_]={x@t,y@t,z@t};
system=MapThread[#1==#2&,{X'[t],A.X[t]}];
sol=DSolve[system,{x,y,z},t]/.{C[1]→-1,C[2]→1,C[3]→1};
plot=ParametricPlot3D[{x@t,y@t,z@t}/.sol,{t,-9,1},
ColorFunction→Function[{x,y,z},Hue[z]],
PlotStyle→{Thick},PlotRange→All];
Show[plot,Graphics3D[{{PointSize[0.03],
ColorFunction→Function[{x,y,z},Hue[z]],
Point[{{x[-9],y[-9],z[-9]}/.sol}]},{PointSize[0.02],
ColorFunction→Function[{x,y,z},Hue[z]],
Point[{{x@1,y@1,z@1}/.sol}]},
{Text[Style["Initial point",Bold,12],{x@-9,y@-9,z@-9+0.5}/.sol],
Text[Style["Final point",Bold,12],{x@1,y@1,z@1+0.5}/.sol]}}],
ImageSize→150]
```

See Fig. 7.5.

Two examples of planar systems of linear homogeneous differential equations
follow. In the first example, the eigenvalues are real and distinct in Fig. 7.6, whereas
in the second example, they are complex in Fig. 7.7, [74, pp. 43–44].

Fig. 7.5 An initial value
problem with a real and two
complex eigenvalues

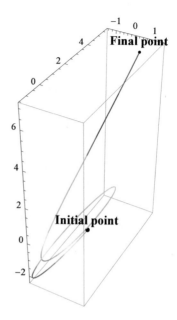

```
solplanarsystemode[A_,x_,y_]:=
Module[{X,system},
X[t_]={x@t,y@t};
system=MapThread[#1==#2&,{X'[t],A.X[t]}];
sol=DSolve[system,{x,y},t]]

Clear[x,y]
A={{4,-6},{1,-1}};
Eigenvalues[A]    (* Real and distinct eigenvalues *)
solplanarsystemode[A,x,y];
particularsols=Partition[Flatten[Table[
{x@t,y@t}/.sol/.{C[1]→1/i,C[2]→1/j},{i,-10,10,6},{j,-10,10,6}]],2];
ParametricPlot[Evaluate[particularsols],{t,-3,3},
PlotRange→{{-2,2},{-1,1}},ImageSize→300]
{2, 1}
```

See Fig. 7.6.

```
Clear[x,y]
A={{7,-8},{5,-5}};
Eigenvalues[A]    (* Complex eigenvalues *)
```

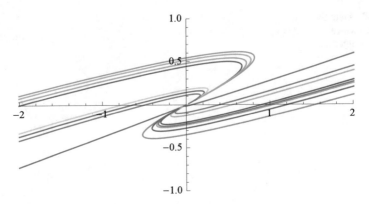

Fig. 7.6 Planar system with real and distinct eigenvalues

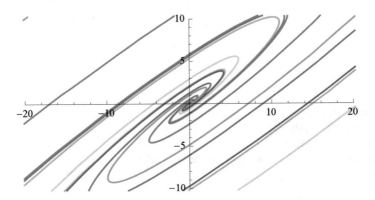

Fig. 7.7 Planar system with complex eigenvalues

```
solplanarsystemode[A,x,y];
particularsols=
Partition[Flatten[Table[
{x@t,y@t}/.sol/.{C[1]→1/i,C[2]→1/j},{i,-10,10,8},{j,-10,10,8}]],2];
ParametricPlot[Evaluate[particularsols],{t,-35,35},
PlotRange→{{-20,20},{-10,10}},ImageSize→300]
{1 + 2i, 1 − 2i}
```
See Fig. 7.7.

7.2.1 Singular Points of the Linear Homogeneous Planar Differential Equations

In some examples we show the behavior of the solutions around their singular points in the case of several linear homogeneous planar differential equations.

Fig. 7.8 Planar system with
null eigenvalues

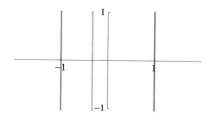

```
Clear[x,y]
A={{0,0},{-3,0}};
Print[" eigenvalues = ",Eigenvalues[A]]    (* Both eigenvalues are null *)
solplanarsystemode[A,x,y];
particularsols=Partition[Flatten[
Table[{x@t,y@t}/.sol/.{C[1]→1/i,C[2]→1/j},{i,-3,2,2},{j,-3,2,2}]],2];
ParametricPlot[Evaluate[particularsols],{t,-10,3},
PlotRange→{{-2,2},{-1,1}},Ticks→{{-1,1},{-1,1}},
ImageSize→200]
eigenvalues = {0, 0}
```

See Fig. 7.8.

```
Clear[x,y]
A={{2,0},{0,2}};
Print[" eigenvalues = ",Eigenvalues[A]]    (* Real positive equal
eigenvalues *)
solplanarsystemode[A,x,y];
particularsols=Partition[Flatten[Table[
{x@t,y@t}/.sol/.{C[1]→1/i,C[2]→1/j},{i,-3,2,2},{j,-3,2,2}]],2];
ParametricPlot[Evaluate[particularsols],{t,-3,3},
PlotRange→{{-1,1},{-1,1}},Ticks→{{-1,1},{-1,1}},
ImageSize→200]
eigenvalues = {2, 2}
```

See Fig. 7.9.

```
Clear[x,y]
A={{1/2,1},{0,1/2}};
Print[" eigenvalues = ",Eigenvalues[A]]    (* Real equal positive
eigenvalues *)
solplanarsystemode[A,x,y];
particularsols=Partition[Flatten[Table[
{x@t,y@t}/.sol/.{C[1]→1/i,C[2]→1/j},{i,-3,2,2},{j,-3,2,2}]],2];
ParametricPlot[Evaluate[particularsols],{t,-15,4},
PlotRange→{{-1.3,1.3},{-1.1,1.1}},Ticks→{{-1,1},{-1,1}},
ImageSize→180]
eigenvalues = {1/2, 1/2}
```

See Fig. 7.10.

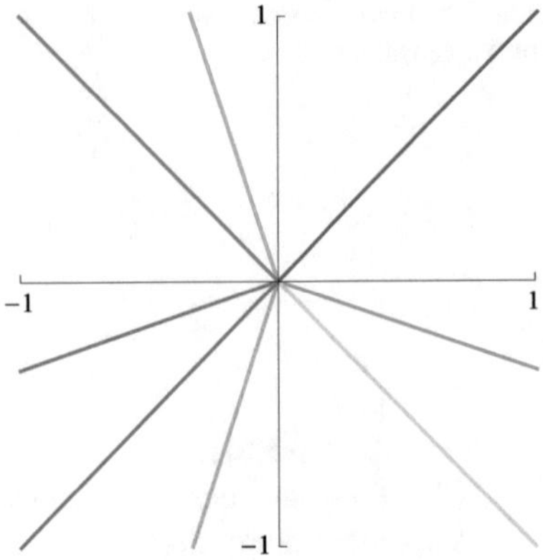

Fig. 7.9 Planar system with
positive equal eigenvalues, 1

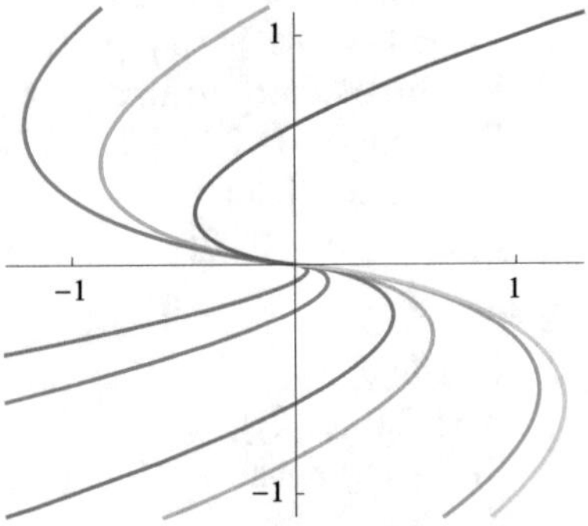

Fig. 7.10 Planar system with
positive equal eigenvalues, 2

```
Clear[x,y]
A={{1/2,0},{0,2}};
Print[" eigenvalues = ",Eigenvalues[A]]    (* Distinct positive eigenvalues *)
solplanarsystemode[A,x,y];
particularsols=Partition[Flatten[Table[
{x@t,y@t}/.sol/.{C[1]→1/i,C[2]→1/j},{i,-3,2,2},{j,-3,2,2}]],2];
ParametricPlot[Evaluate[particularsols],{t,-15,3},
PlotRange→{{-1.3,1.3},{-1,1}},Ticks→{{-1,1},{-1,1}},
ImageSize→180]
eigenvalues = {2, 1/2}
```

See Fig. 7.11.

```
Clear[x,y]
A={{0,1},{1,0}};
Print[" eigenvalues = ",Eigenvalues[A]]    (* Distinct eigenvalues,
one positive, one negative *)
solplanarsystemode[A,x,y];
particularsols=Partition[Flatten[Table[
{x@t,y@t}/.sol/.{C[1]→1/i,C[2]→1/j},{i,-3,2,1.2},{j,-3,2,1.2}]],2];
```

Fig. 7.11 Planar system with
distinct positive eigenvalues

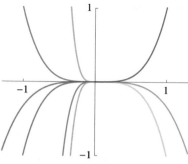

Fig. 7.12 Planar system with
one positive and one negative
eigenvalues

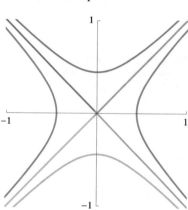

```
ParametricPlot[Evaluate[particularsols],{t,-4,4},
PlotRange→{{-1,1},{-1,1}},Ticks→{{-1,1},{-1,1}},
ImageSize→180]
eigenvalues = {−1, 1}
```

See Fig. 7.12.

```
Clear[x,y]
A={{-2,0},{0,-2}};
Print["eigenvalues = ",Eigenvalues[A]]   (* Real equal negative
eigenvalues *)
solplanarsystemode[A,x,y];
particularsols=Partition[Flatten[Table[
{x@t,y@t}/.sol/.{C[1]→1/i,C[2]→1/j},{i,-3,2,1.2},{j,-3,2,1.2}]],2];
ParametricPlot[Evaluate[particularsols],{t,-3,3},
PlotRange→{{-1,1},{-1,1}},Ticks→{{-1,1},{-1,1}},
ImageSize→180]
eigenvalues = {−2, −2}
```

See Fig. 7.13.

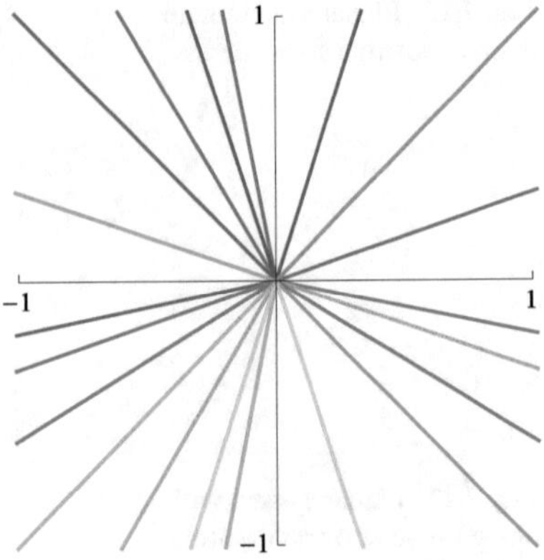

Fig. 7.13 Planar system with
equal negative eigenvalues

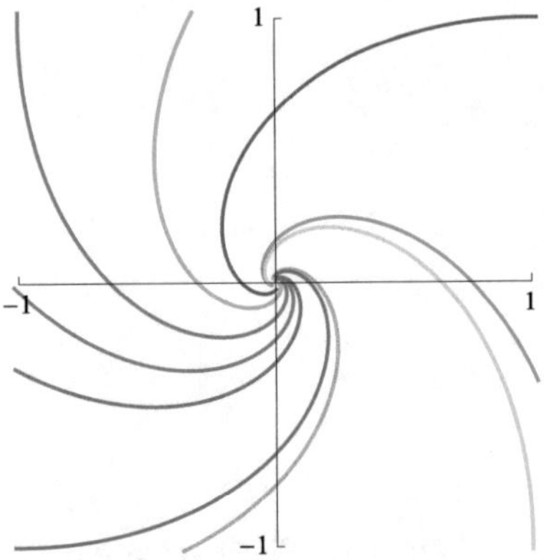

Fig. 7.14 Planar system with
complex distinct eigenvalues
with negative real parts

```
Clear[x,y]
A={{-1,-1},{1,-1}};
Print[" eigenvalues = ",Eigenvalues[A]]    (* Complex distinct eigenvalues
with negative real parts *)
solplanarsystemode[A,x,y];
particularsols=Partition[Flatten[Table[
{x@t,y@t}/.sol/.{C[1]→1/i,C[2]→1/j},{i,-3,2,2},{j,-3,2,2}]],2];
ParametricPlot[Evaluate[particularsols],{t,-10,4},
PlotRange→{{-1,1},{-1,1}},Ticks→{{-1,1},{-1,1}},
ImageSize→180]
```
eigenvalues $= \{-1 + i, -1 - i\}$

See Fig. 7.14.

Fig. 7.15 Planar system with complex distinct eigenvalues with positive real parts

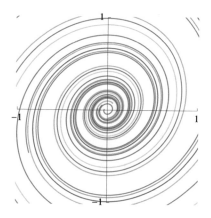

```
Clear[x,y]
A={{1/2,-2},{2,0}};
Print[" eigenvalues = ",Eigenvalues[A]]    (* Complex distinct
eigenvalues, the real part is positive *)
solplanarsystemode[A,x,y];
particularsols=Partition[Flatten[Table[
{x@t,y@t}/.sol/.{C[1]→1/i,C[2]→1/j},{i,-3,2,2},{j,-3,2,2}]],2];
ParametricPlot[Evaluate[particularsols],{t,-10,3},
PlotRange→{{-1,1},{-1,1}},Ticks→{{-1,1},{-1,1}},
ImageSize→180]
```
eigenvalues $= \left\{ \frac{1}{4}\left(1 + 3i\sqrt{7}\right), \frac{1}{4}\left(1 - 3i\sqrt{7}\right) \right\}$

See Fig. 7.15.

```
Clear[x,y]
A={{0,-1},{1,0}};
Print[" eigenvalues = ",Eigenvalues[A]]    (* Pure complex eigenvalues *)
solplanarsystemode[A,x,y];
particularsols=Partition[Flatten[
Table[{x@t,y@t}/.sol/.{C[1]→1/i,C[2]→1/j},{i,-3,2,1.2},{j,-3,2,1.2}]],2];
ParametricPlot[Evaluate[particularsols],{t,-4,4},
PlotRange→{{-1,1},{-1,1}},Ticks→{{-1,1},{-1,1}},
ImageSize→200]
```
eigenvalues $= \{i, -i\}$

See Fig. 7.16.

  We collect all the above results in a single code.

Fig. 7.16 Planar system with
pure complex eigenvalues

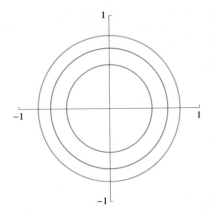

```
Clear[x,y]
mat={{{0,0},{-3,0}},{{2,0},{0,2}},{{1/2.,1},{0,1/2}},
{{1./2,0},{0,2}},{{0,1},{1,0}},{{-2,0},{0,-2}},{{-1,-1},{1,-1}},
{{1,-3},{3,1}},{{0,-1},{1,0}}};
colour={Red,Blue,Magenta,Green,Darker[Red],Brown,Blue,
Darker[Green],Red};
font={"Helvetica","Times","Courier","Old English Text MT"};
Table[A=mat[[i]];
eigen=Eigenvalues[A];
solplanarsystemode[A,x,y];
particularsols=Partition[Flatten[Table[
{x@t,y@t}/.sol/.{C[1]→1/m,C[2]→1/j},{m,-3,2,1.2},{j,-3,2,1.2}]],2];
ParametricPlot[Evaluate[particularsols],{t,-8,3},
PlotLabel→ToExpression["eigenvalues"]==TraditionalForm[eigen],
LabelStyle→{colour[[i]],FontFamily→font[[Mod[i,4]+1]]},
PlotRange→{{-1.001,1},{-1,1.1}},Ticks→{{-1,1},{-1,1}},
PlotStyle→{Directive[Thickness→0.008,color[[i]]]}],{i,1,9}]
```

See Fig. 7.17.

7.2.2 Multiple Equilibria

7.2.2.1 Tunnel-Diode Circuit

Consider the following system of planar nonlinear ordinary differential equations

$$\begin{cases} x' = 0.5(-h(x) + y), \\ y' = 0.2(-x - 1.5y + 1.2), \end{cases}$$

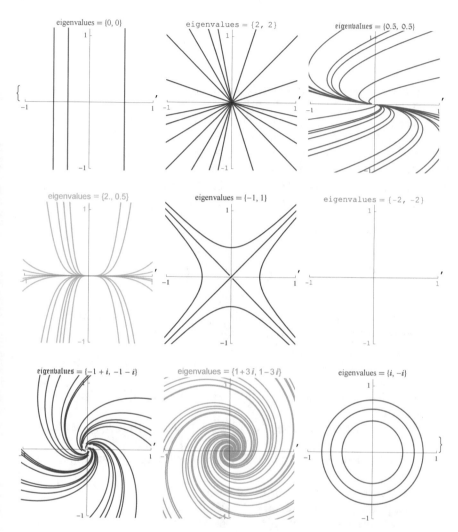

Fig. 7.17 All graphs of the earlier planar systems

where $h(x) = 17.76x - 103.79x^2 + 229.62x^3 - 226.31x^4 + 83.72x^5$. The *equilibrium points* of this system are the real solutions of the system

$$\begin{cases} 0.5(-h(x) + y) = 0, \\ -x - 1.5y + 1.2 = 0. \end{cases}$$

Let P_1, P_2, and P_3 be the equilibrium points. The first picture shows the three equilibrium points. The second picture shows the trajectories of the system in the phase-plane near the equilibrium points.

```
Clear[x,y,h,p,A]
h[x_]=17.76x-103.79x²+229.62x³-226.31x⁴+83.72x⁵;
A={0.5{-h[x],y},0.2{-x,-1.5 y+1.2}};
diffA={{D[A[[1,1]],x],D[A[[1,2]],y]},{D[A[[2,1]],x],D[A[[2,2]],y]}};
soll=Solve[h[x]-y==0&&-x-1.5y+1.2==0,{x,y},Reals];
Table[p[i]={x,y}/.soll[[i]],{i,1,3}];
Table[Eigenvalues[diffA/.{x→p[i][[1]],y→p[i][[2]]}],{i,1,3}];
{Show[Plot[(1.2-x)/1.5,{x,0,1.2},
PlotStyle→{Directive[Thickness→0.005,Blue]}],
Plot[h[x],{x,0,1.2},PlotRange→{{0,1.25},{0,1}},
PlotStyle→{Directive[Thickness→0.005,Blue]}],
Graphics[{{PointSize[0.017],Red,Point[p[1]],Point[p[2]],Point[p[3]]},
{Text[Style[" P₁" ,Italic,12],p[1]+{0.068,0.04}],
Text[Style[" y=h(x)" ,Italic,12],p[3]+{-.028,.67}],
Text[Style[" x+1.5y=1.2" ,Italic,12],{.67,.56}],
Text[Style[" P₂" ,Italic,12],p[2]+{0.06,0.04}],
Text[Style[" P₃" ,Italic,12],p[3]+{0.07,0.04}]}}],
PlotRange→All,ImageSize→{250,180}],
eqs=Sequence[u'[t]==0.5(-h[u[t]]+v[t]),v'[t]==0.2(-u[t]-1.5 v[t]+1.2)];
Show[Table[sol2=NDSolve[{eqs,u[0]==p[1][[1]]+i/4,
v[0]==p[1][[2]]+j/4},{u,v},{t,0,100}];
sol3=NDSolve[{eqs,u[0]==p[2][[1]]+i/4,v[0]==p[2][[2]]+j/4},{u,v},
{t,0,100}];
sol4=NDSolve[{eqs,u[0]==p[3][[1]]+i/4,v[0]==p[3][[2]]+j/4},{u,v},
{t,0,100}];
ParametricPlot[{Evaluate[{u[t],v[t]}/.sol2],Evaluate[{u[t],v[t]}/.sol3],
Evaluate[{u[t],v[t]}/.sol4]},{t,0,100}],{i,-2,2},{j,-2,2}],
Graphics[{PointSize[0.015],Red,Point[p[1]],Point[p[2]],Point[p[3]],
Black,
Text[Style[" P₁" ,Italic,12],p[1]+{0.068,0}],
Text[Style[" P₂" ,Italic,12],p[2]+{-0.065,0}],
Text[Style[" P₃" ,Italic,12],p[3]+{0.06,-0.065}]}],
PlotRange→{{-.4,1.4},{-.3,1.3}},AxesOrigin→{0,0},
ImageSize→{320,300}]}
```

See Fig. 7.18.

7.2.2.2 Five Equilibrium Points

Here we introduce a planar system of ordinary differential equations with 5
equilibrium points. We introduce two codes.

In the first code we study certain trajectories in order to clearly see the phase
portrait. The second code is much shorter.

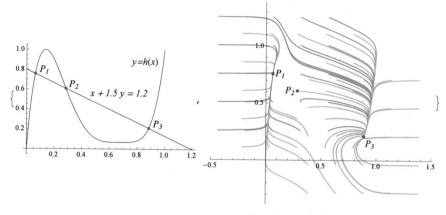

Fig. 7.18 Tunnel-diode circuit

The system under discussion is given below:

$$\begin{cases} x' = -y - x^3/2 + x^4/2 + 3y^4/2 - 3y^5/4, \\ y' = 4x - 6y^3 + x^4 + 3y^4 - x^5. \end{cases}$$

Here is the first code:

```
Clear[x,y]
equipointreal=N[Simplify[Solve[-y-x³/2+x⁴/2+3y⁴/2-3y⁵/4==0&&
4x-6y³+x⁴+3y⁴-x⁵==0,{x,y},Reals]]];
equipointreal={x,y}/.equipointreal;

listofpoints[m_,h_]:=
Module[{k,u,v},
For[k=0,k<m,k++,
u = x + h(−y − x³/2 + x⁴/2 + 3y⁴/2 − 3y⁵/4);
v = y + h(4x − 6y³ + x⁴ + 3y⁴ − x⁵);
AppendTo[points,{u,v}];x=u;y=v]]

points={equipointreal[[5]]};
{x,y}=equipointreal[[5]]+{-0.01,0.01};
h=0.02;nsteps=8;listofpoints[nsteps,h];
arrow5nw={points[[Floor[nsteps/2]+3]],points[[Floor[nsteps/2]+4]]};
plot5nw=ListLinePlot@N@points;

points={equipointreal[[5]]};
{x,y}=equipointreal[[5]]+{0.01,-0.01};
h=0.02;nsteps=14;listofpoints[nsteps,h];
arrow5se={points[[Floor[nsteps/2]+2]],points[[Floor[nsteps/2]+3]]};
plot5se=ListLinePlot@N@points;
```

```
points={equipointreal[[5]]};
{x,y}=equipointreal[[5]]+{-0.01,-0.01};
h=0.02;nsteps=150;listofpoints[nsteps,-h];
arrow5sw={points[[Floor[nsteps/3]+4]],points[[Floor[nsteps/3]+3]]};
plot5sw=ListLinePlot@N@points;

points={equipointreal[[5]]};
{x,y}=equipointreal[[5]]+{0.01,0.01};
h=0.02;nsteps=48;listofpoints[nsteps,-h];
arrow5ne={points[[Floor[nsteps/2]+5]],points[[Floor[nsteps/2]+4]]};
plot5ne=ListLinePlot[N[points]];

points={equipointreal[[4]]};
{x,y}=equipointreal[[4]]+{-0.01,0.01};
h=0.02;nsteps=9;listofpoints[nsteps,h];
arrow4nw={points[[Floor[nsteps/2]+4]],points[[Floor[nsteps/2]+5]]};
plot4nw=ListLinePlot@N@points;

points={equipointreal[[4]]};
{x,y}=equipointreal[[4]]+{0.00001,-0.00001};
h=0.02;nsteps=265;listofpoints[nsteps,h];
arrow4se={points[[Floor[nsteps/11]+2]],points[[Floor[nsteps/11]+5]]};
plot4se=ListLinePlot[N[points]];

points={equipointreal[[4]]};
{x,y}=equipointreal[[4]]+{-0.01,-0.01};
h=0.02;nsteps=29;listofpoints[nsteps,-h];
arrow4sw={points[[Floor[nsteps/2]+3]],points[[Floor[nsteps/2]+2]]};
plot4sw=ListLinePlot@N@points;

points={equipointreal[[4]]};
{x,y}=equipointreal[[4]]+{0.01,0.0001};
h=0.02;nsteps=100;listofpoints[nsteps,-h];
arrow4e={points[[Floor[nsteps/5]+3]],points[[Floor[nsteps/5]+2]]};
plot4e=ListLinePlot@N@points;

points={points[[Floor[nsteps/2]]]};
{x,y}=Flatten[points]-{0,0.001};
h=0.02;nsteps=100;listofpoints[nsteps,h];
arrow3ws={points[[Floor[nsteps/4]+1]],points[[Floor[nsteps/4]+2]]};
plot3ws=ListLinePlot@N@points;
```

```
points={equipointreal[[3]]};
{x,y}=equipointreal[[3]]+{-0.01,0.01};
h=0.02;nsteps=30;listofpoints[nsteps,h];
arrow3nw={points[[Floor[nsteps/2]+9]],points[[Floor[nsteps/2]+10]]};
plot3nw=ListLinePlot@N@points;

points={equipointreal[[3]]};
{x,y}=equipointreal[[3]]+{-0.1,0.001};
h=0.02;nsteps=50;listofpoints[nsteps,h];
arrow3w={points[[Floor[nsteps/2]+2]],points[[Floor[nsteps/2]+3]]};
plot3w=ListLinePlot@N@points;

points={equipointreal[[2]]};
{x,y}=equipointreal[[2]]+{0.01,0.005};
h=0.02;nsteps=173;listofpoints[nsteps,h];
arrow2e={points[[Floor[nsteps/2]+60]],points[[Floor[nsteps/2]+61]]};
plot2e=ListLinePlot[N[points],InterpolationOrder→2];

points={equipointreal[[2]]};
{x,y}=equipointreal[[2]]+{-0.01,-0.001};
h=0.02;nsteps=1000;listofpoints[nsteps,h];
arrow2sw={points[[Floor[nsteps/6]+24]],points[[Floor[nsteps/6]+25]]};
plot2sw=ListLinePlot@N@points;

points={equipointreal[[2]]};
{x,y}=equipointreal[[2]]+{-0.001,0.001};
h=0.02;nsteps=150;listofpoints[nsteps,-h];
arrow2n={points[[Floor[nsteps/4]+9]],points[[Floor[nsteps/4]+8]]};
plot2n=ListLinePlot@N@points;

points={equipointreal[[2]]};
{x,y}=equipointreal[[2]]+{0.01,-0.01};
h=0.02;nsteps=74;listofpoints[nsteps,-h];
arrow2s={points[[Floor[nsteps/3]+9]],points[[Floor[nsteps/3]+8]]};
plot2s=ListLinePlot@N@points;

{x,y}={2.7,2.65};
points={{x,y}};
h=0.002;nsteps=18;listofpoints[nsteps,h];
arrow5fs={points[[Floor[nsteps/2]]],points[[Floor[nsteps/2]+1]]};
plot5fs=ListLinePlot[N@points,InterpolationOrder→2];

{x,y}={2.7,2.9};
points={{x,y}};
h=0.002;nsteps=9;listofpoints[nsteps,h];
```

```
arrow5fn={points[[Floor[nsteps/2]+3]],points[[Floor[nsteps/2]+4]]};
plot5fn=ListLinePlot[N@points,InterpolationOrder→2];
h=0.02;

{x,y}={-2,-1.5};
points={{x,y}};
h=0.02;nsteps=200;listofpoints[nsteps,h];
arrow1fn={points[[Floor[nsteps/5]+10]],points[[Floor[nsteps/5]+11]]};
plot1fn=ListLinePlot[N[points],InterpolationOrder→4];

{x,y}={-2.7,-1.1};
points={{x,y}};
h=0.002;nsteps=32;listofpoints[nsteps,h];
arrowfsn={points[[Floor[nsteps/2]+1]],points[[Floor[nsteps/2]+2]]};
plotfsn=ListLinePlot[N[points]];

Show[plot5nw,plot5se,plot5ne,plot5sw,plot4nw,plot4se,plot4sw,
plot4e,plot3nw,plot2e,plot2sw,plot2n,plot2s,plot5fs,plot5fn,plot1fn,
plotfsn,plot3ws,
Graphics[{{PointSize[0.013],Red,Point[{equipointreal[[5]],
equipointreal[[4]],equipointreal[[3]],equipointreal[[2]],
equipointreal[[1]]}]}},
Arrowheads[0.028],Green,
Arrow[{arrow5nw,arrow5sw,arrow5ne,arrow5se,arrow4nw,arrow4se,
arrow4sw,arrow4e,arrow3nw,arrow2e,arrow2sw,arrow2n,arrow2s,
arrow5fs,arrow5fn,arrow1fn,arrowfsn,arrow3ws}]}],
Ticks→{{-3,-2,-1,1,2,3},{-3,-2,-1,1,2,3,4}},AxesOrigin→{0,0},
PlotPoints→40,AspectRatio→1,ImageSize→300,PlotRange→All]
```

See Fig. 7.19.

The second code follows:

```
Clear[x,y]
equipoints=Solve[-y-x³/2+x⁴/2+3y⁴/2-3y⁵/4==0&&
4x-6y³+x⁴+3y⁴-x⁵==0,{x,y},Reals];
eqreal={x,y}/.equipoints;

vector={{eqreal[[5]]+{-0.01,0.01},eqreal[[5]]+{0.01,-0.01},
eqreal[[5]]+{-0.01,-0.01},eqreal[[5]]+{0.01,0.01},
eqreal[[5]]+{0.01,-0.01},eqreal[[5]]+{-0.01,-0.01},
eqreal[[5]]+{0.01,0.01},eqreal[[4]]+{-0.01,0.01},}
eqreal[[4]]+{0.00001,-0.00001},eqreal[[4]]+{-0.01,-0.01},
eqreal[[4]]+{0.01,0.0001},eqreal[[3]]+{-0.01,0.01},
eqreal[[2]]+{0.01,0.005},eqreal[[2]]+{-0.01,-0.001},
eqreal[[2]]+{-0.001,0.001},eqreal[[2]]+{0.01,-0.01},
{2.7,2.65},{2.7,2.9},{-2,-1.5},{-2.7,-1.1}},
{8,14,150,48,14,150,48,9,265,29,100,30,173,1000,150,74,18,9,200,32},
```

Fig. 7.19 First code for the
five equilibrium case

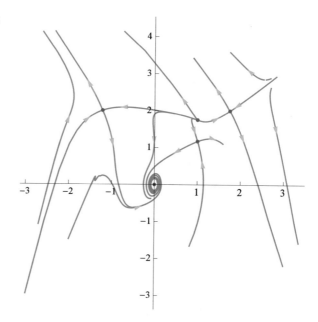

```
{0.02,0.02,-0.02,-0.02,0.02,-0.02,-0.02,0.02,0.02,-0.02,-0.02,0.02,
0.02,0.02,-0.02,-0.02,0.002,0.002,0.02,0.002},
{2,2,3,2, 2,3,2,2,11,2,5,2, 2,6,4,3,2,2,5,2},
{{3,4},{2,3},{4,3},{5,4},{2,3},{4,3},{5,4},{4,5},{2,5},{3,2},{3,2},
{9,10},{60,61},{24,25},{9,8},{9,8},{0,1},{3,4},{10,11},{1,2}}};
aArrows=aPlots={};
Do[{x,y}=Flatten[vector[[1,m]]];points={{x,y}};
nsteps=vector[[2,m]];
For[k=0,k<nsteps,k++,
u=x+vector[[3,m]](-y-x³/2+x⁴/2+3y⁴/2-3y⁵/4);
v=y+vector[[3,m]](4x-6y³+x⁴+3y⁴-x⁵);
AppendTo[points,{u,v}];x=u;y=v];
aArrows=AppendTo[aArrows,
{points[[Floor[nsteps/vector[[4,m]]+vector[[5,m,1]]]]],
points[[Floor[nsteps/vector[[4,m]]+vector[[5,m,2]]]]]}];
aPlots=AppendTo[aPlots,ListLinePlot[N@points]],
{m,Length[vector[[1]]]}]
Show[Flatten[aPlots],
Graphics[{{PointSize[0.013],Red,
Point[{eqreal[[5]],eqreal[[4]],eqreal[[3]],eqreal[[2]],eqreal[[1]]}]},
Arrowheads[0.028],Green,Arrow[aArrows]}],AxesOrigin→{0,0},
AspectRatio→1,Ticks→{{-3,-2,-1,1,2,3},{-3,-2,-1,1,2,3,4}},
PlotRange→All]
```

See Fig. 7.20.

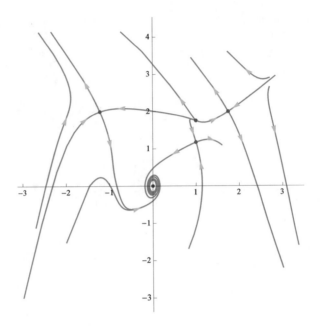

Fig. 7.20 Second code for the five equilibrium case

7.3   On Two Runge–Kutta Methods of the Fourth Order

In this section we will introduce an explicit fourth-order Runge–Kutta method, which requires only three evaluations at each step, and a semi-explicit Runge–Kutta method of fourth order with two evaluations at each step. These methods were worked out in [49, 50], and [51].

7.3.1   An Explicit Runge–Kutta Method

For the explicit fourth-order Runge–Kutta method which requires only three evaluations at each step, we consider a system of autonomous ordinary differential equations of the first order in the vectorial form

$$\frac{dx}{dt} = f(x) \tag{7.1}$$

where $x = (x_1, \ldots, x_s)$, $f = f(x_1, \ldots, x_s)$, $t \in \mathbb{R}$.

We suppose that the vector $x = x_0$ is given at $t = a$. The interval $[a, b]$, where the independent variable t belongs, is partitioned by a finite mesh $h_i > 0$ and

$$a + \sum_{i=1}^{m} h_i = b. \tag{7.2}$$

We also suppose that the function f is sufficiently smooth such that all that follow are correct. This implies that the initial value problem stated at $x(a) = x_0$ has a unique solution which exists on the whole interval $[a, b]$. The sequence of approximations (x_n) of the unknown function x solution of the initial value problem consisting in the system of differential equations (7.1) and the initial condition is defined by

$$x_{n+1} = x_n + h_n \left(\alpha_0 k_{0,n} + \alpha_1 k_{1,n} + \alpha_2 k_{2,n} \right),$$

$$k_{i,n} = f \left(\arg_{i,n} \right),$$

$$\arg_{i,n} = x_n + h_n \left(\sum_{j=0}^{2} \lambda_{i,j} k_{j,n-1} + \rho_{i,0} k_{0,n} + \rho_{i,1} k_{1,n} \right),$$

$$i = 0, 1, 2 \quad \text{and} \quad n = 1, 2, \ldots, m.$$

For the first step, we take

$$k_{i,-1} = x_0, \quad i = 0, 1, 2.$$

The method is explicit, so we have

$$\rho_{0,0} = \rho_{0,1} = \rho_{1,1} = 0.$$

To get an accuracy up to the terms in h^4, the following relations concerning the parameters involved have to hold

$$\lambda_{0,0} + \lambda_{0,1} + \lambda_{0,2} = \mu_0,$$
$$\lambda_{1,0} + \lambda_{1,1} + \lambda_{1,2} + \rho_{1,0} = \mu_1,$$
$$\lambda_{2,0} + \lambda_{2,1} + \lambda_{2,2} + \rho_{2,0} + \rho_{2,1} = \mu_2,$$
$$\alpha_0 + \alpha_1 + \alpha_2 = 1,$$
$$2 \left(\alpha_0 \mu_0 + \alpha_1 \mu_1 + \alpha_2 \mu_2 \right) = 1,$$
$$3 \left(\alpha_0 \mu_0^2 + \alpha_1 \mu_1^2 + \alpha_2 \mu_2^2 \right) = 1,$$
$$4 \left(\alpha_0 \mu_0^3 + \alpha_1 \mu_1^3 + \alpha_2 \mu_2^3 \right) = 1,$$
$$6 \left(\alpha_1 \rho_{1,0} \mu_0 + \alpha_2 \left(\rho_{2,0} \mu_0 + \rho_{2,1} \mu_1 \right) \right) = 1,$$
$$8 \left(\alpha_1 \rho_{1,0} \mu_0 \mu_1 + \alpha_2 \mu_2 \left(\rho_{2,0} \mu_0 + \rho_{2,1} \mu_1 \right) \right) = 1,$$
$$12 \left(\alpha_1 \rho_{1,0} \mu_0^2 + \alpha_2 \left(\rho_{2,0} \mu_0^2 + \rho_{2,1} \mu_1^2 \right) \right) = 1,$$

$$24\alpha_2\rho_{1,0}\rho_{2,1}\mu_0 = 1,$$
$$\lambda_{0,0}\mu_0 + \lambda_{0,1}\mu_1 + \lambda_{0,2}\mu_2 = \mu_0,$$
$$\lambda_{1,0}\mu_0 + \lambda_{1,1}\mu_1 + \lambda_{1,2}\mu_2 + \rho_{1,0} = \mu_1,$$
$$\lambda_{2,0}\mu_0 + \lambda_{2,1}\mu_1 + \lambda_{2,2}\mu_2 + \rho_{2,0} + \rho_{2,1} = \mu_2,$$
$$2\left(\sum_{i=0}^2 \alpha_i \sum_{k=0}^2 \lambda_{i,k}\mu_k + sS\right) = 1,$$
$$2\left(\sum_{i=0}^2 \alpha_i \sum_{k=0}^2 \lambda_{i,k}\mu_k^2 + sS\right) = 1,$$
$$2\left(\sum_{i=0}^2 \alpha_i \left(\lambda_{i,1}\rho_{1,0}\mu_0 + \lambda_{i,2}\left(\rho_{2,0}\mu_0 + \rho_{2,1}\mu_1\right)\right) + sS\right) = 1,$$

where

$$sS = \alpha_1\rho_{1,0} + \alpha_2\left(\rho_{2,0} + \rho_{2,1}\right).$$

In [49] we have expressed these parameters in terms of μ_0, μ_1, μ_2, $\lambda_{1,1}$, and $\lambda_{2,1}$. From the compatibility condition, we get $\mu_2 = 1$ and the equality

$$6\mu_0\mu_1 - 2\left(\mu_0 + \mu_1\right) + 1 = 0.$$

Successively we have the results

$$\alpha_0 = \frac{3\mu_1 - 1}{6\left(1 - \mu_0\right)\left(\mu_1 - \mu_0\right)}, \qquad \rho_{1,0} = \frac{\mu_1 - \mu_0}{4\mu_0\left(1 - 3\mu_0\right)},$$

$$\alpha_1 = \frac{3\mu_0 - 1}{6\left(\mu_1 - 1\right)\left(\mu_1 - \mu_0\right)}, \qquad \rho_{2,0} = \frac{\left(1 - \mu_0\right)\left(4\mu_1 - 1\right)\left(3\mu_0 - \mu_1\right)}{2\mu_0\left(\mu_1 - \mu_0\right)\left(1 - 6\mu_0\mu_1\right)},$$

$$\alpha_2 = \frac{1 - 6\mu_0\mu_1}{12\left(1 - \mu_1\right)\left(1 - \mu_0\right)}, \qquad \rho_{2,1} = \frac{\left(1 - 3\mu_0\right)\left(1 - 4\mu_0\mu_1\right)}{\left(\mu_1 - \mu_0\right)\left(1 - 6\mu_0\mu_1\right)},$$

$$sS = 1/2.$$

Further, if we denote

$$\lambda_{1,1} = \beta, \quad \lambda_{2,1} = \gamma, \text{ and } \delta_0 = \frac{\mu_1 - 1}{\mu_0 - 1},$$

it results

$$\lambda_{1,0} = -\delta\beta, \quad \lambda_{2,0} = -\delta\gamma, \quad \lambda_{0,1} = -\frac{\alpha_1\beta + \alpha_2\delta}{\alpha_0},$$

$$\lambda_{0,0} = \frac{-\delta\left(\alpha_1\beta + \alpha_2\delta\right)}{\alpha_0}, \quad \lambda_{0,2} = \mu_0 - \frac{\left(\delta - 1\right)\left(\alpha_1\beta + \alpha_2\delta\right)}{\alpha_0},$$

$$\lambda_{1,2} = \mu_1 - \rho_{1,0} + \left(\delta - 1\right)\beta, \quad \lambda_{2,2} = \mu_2 - \left(\rho_{2,0} + \rho_{2,1}\right) + \left(\delta - 1\right)\gamma,$$

Then we immediately have

$$\alpha_0 = \frac{3\mu_1 - 1}{6(1-\mu_0)(\mu_1 - \mu_0)};$$
$$\alpha_1 = \frac{3\mu_0 - 1}{6(\mu_1 - 1)(\mu_1 - \mu_0)};$$
$$\alpha_2 = \frac{1 - 6\mu_0\mu_1}{12(1-\mu_1)(1-\mu_0)};$$
$$(*\rho_{0,0} = \rho_{0,1} = \rho_{1,1} = 0; *)$$
$$\rho_{1,0} = \frac{\mu_1 - \mu_0}{4\mu_0(1-3\mu_0)};$$
$$\rho_{2,0} = \frac{(1-\mu_0)(4\mu_1 - 1)(3\mu_0 - \mu_1)}{2\mu_0(\mu_1 - \mu_0)(1-6\mu_0\mu_1)};$$
$$\rho_{2,1} = \frac{(1-3\mu_0)(1-4\mu_0\mu_1)}{(\mu_1 - \mu_0)(1-6\mu_0\mu_1)};$$
$$\lambda_{1,1} = \mu_1 - (\rho_{1,0} + \mu_0\lambda_{1,0} + \lambda_{2,1})$$
$$\lambda_{1,1} = \beta = 0.0;$$
$$\lambda_{2,1} = \gamma = 0.0;$$
$$\delta = \frac{\mu_1 - 1}{\mu_0 - 1};$$
$$\lambda_{0,0} = -\delta \times \frac{\alpha_1\beta + \alpha_2\gamma}{\alpha_0};$$
$$\lambda_{0,1} = -\frac{\alpha_1\beta + \alpha_2\gamma}{\alpha_0};$$
$$\lambda_{0,2} = \mu_0 - (\delta - 1)\frac{\alpha_1\beta + \alpha_2\gamma}{\alpha_0};$$

$$\lambda_{1,0} = -\delta\,\beta;$$
$$\lambda_{1,2} = \mu_1 - \rho_{1,0} + (\delta - 1)\beta;$$
$$\lambda_{2,0} = -\delta\,\gamma;$$
$$\lambda_{2,2} = \mu_2 - (\rho_{2,0} + \rho_{2,1}) + (\delta - 1)\gamma;$$
$$\mu_1 - \frac{-\mu_0 + \mu_1}{4(1-3\mu_0)\mu_0} - \mu_0\lambda_{1,0} - \lambda_{2,1}$$

By some numerical experiments it seems reasonable to consider that

$$\mu_0 = 0.5, \quad \mu_1 = 0.0, \quad \mu_2 = 1.0, \quad \lambda_{1,1} = \mu_1 - (\rho_{1,0} + \mu_0\lambda_{1,0} + \lambda_{2,1}).$$

We consider the next example. The approach by our Runge–Kutta method follows.

```
f[{x_,y_,z_}]:={1,Cos[x]²y-(1-Sin[x]Cos[x])z,(1+Sin[x]Cos[x])y+Sin[x]²z}
(* The right-hand side of the system of differential equations *)
x0={0,1,1};   (* Initial value *)
a=0;   (* Leftmost boundary value *)
b=π;   (* Rightmost boundary value *)
m=1000;   (* Number of steps *)
h=(b-a)/m;   (* Step size uniformly distributed *)
kold0=kold1=kold2=x0;
Do[
knew0=f[x0+h(λ0,0kold0+λ0,1kold1+λ0,2kold2)];
knew1=f[x0+h(λ1,0kold0+λ1,1kold1+λ1,2kold2+ρ1,0knew0)];
knew2=f[x0+h(λ2,0kold0+λ2,1kold1+λ2,2kold2+ρ2,0knew0+ρ2,1knew1)];
x1=x0+h(α0knew0+α1knew1+α2knew2);
x0=x1;
kold0=knew0;kold1=knew1;kold2=knew2,
```

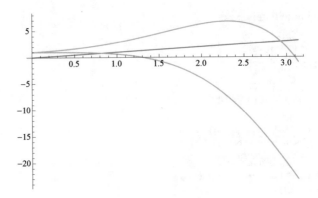

Fig. 7.21 The graphs of the exact and approximative solutions

```
{i,m}]
N[x1,15]
{3.14159,-23.1405,-0.989255}
```

The solution by the built-in function NDSolve is presented below.

```
Clear[x,y,z,t]
sol=NDSolve[{x'[t]==1,y'[t]==Cos[x[t]]²y[t]-(1-Sin[x[t]]Cos[x[t]])z[t],
z'[t]==(1+Sin[x[t]]Cos[x[t]])y[t]+Sin[x[t]]²z[t],x[0]==0,y[0]==1,z[0]==1},
{x,y,z},{t,π},AccuracyGoal→20,PrecisionGoal→20,
WorkingPrecision→35];
N[{x[π],y[π],z[π]}/.sol,15]    (* The solution by integration *)
N[{t,Exp[t]Cos[t]-Sin[t],Exp[t]Sin[t]+Cos[t]}/.t→ π,15]    (* The exact
solution at π *)
{{3.14159265358979,-23.1406926327793,-1.00000000000000}}
{3.14159265358979,-23.1406926327793,-1.00000000000000}

Plot[{t,Exp[t]Cos[t]-Sin[t],Exp[t]Sin[t]+Cos[t]},{t,0,π},PlotRange→All]
(* The graph of the exact solution *)
```
See Fig. 7.21.

7.3.2 A Semi-explicit Runge–Kutta Method

We consider a system of first-order autonomous ordinary differential equations in the vectorial form (7.1) where $x = (x_1, \ldots, x_s), f = f(x_1, \ldots, x_s), t \in [a, b]$. We admit that a vector is given $x_0 = x(a) \in \mathbb{R}^s$ at $t = a$. The interval $[a, b]$, where the independent variable t belongs, is partitioned by a finite mesh $h_i > 0$ and (7.2) holds.

We also suppose that the function f is sufficiently smooth such that all that follow are correct. This implies that the initial value problem stated at $x(a) = x_0$

has a unique solution, which exists on the whole interval $[a, b]$. The sequence of approximations (x_n) of the unknown function x solution of the initial value problem consisting in the system of differential equations (7.1) and the initial condition is well defined.

It is known that to each initial value problem, we assign a table of coefficients of the form

Clear[a,b,τ]

Grid[{{$a_{1,1},\ldots,a_{1,q},\tau_1$},{:, :::},{$a_{q,1},\ldots,a_{q,q},\tau_q$},{$b_1,\ldots,b_q$,}},
Dividers→{{False,False,False,True},{False,False,False,True}}]

$$
\begin{array}{ccc|c}
a_{1,1} & \cdots & a_{1,q} & \tau_1 \\
\vdots & & \vdots & \vdots \\
a_{q,1} & \cdots & a_{q,q} & \tau_q \\
\hline
b_1 & \cdots & b_q &
\end{array}
$$

where

$$
\sum_{i=1}^{q} a_{j,i} = \tau_j,
$$

$$
x_{n+1} = x_n + h_n \sum_{i=1}^{q} b_i k_{i,n}, \tag{7.3}
$$

$$
k_{i,n} = f\left(\arg_{i,n}\right), \quad \arg_{i,n} = x_n + h_n \sum_{j=1}^{q} a_{i,j} k_{j,n}.
$$

A Runge–Kutta method is said to be *explicit* if $a_{i,j} = 0$ for $i \le j$; is said to be *semi-explicit* if $a_{i,j} = 0$ for $i < j$; is *diagonally implicit* if is semi-explicit and all the $a_{i,i}$ are equal, [4].

We denote by p the order of the method. In [4] Alexander shows that there exists a unique diagonally implicit formula with $(q, p) = (1, 2)$, as the implicit midpoint rule:

Grid[{{1/2,1/2},{1,Null}},Dividers→{{False,True},{False,True}}]

$$
\begin{array}{c|c}
\frac{1}{2} & \frac{1}{2} \\
\hline
1 &
\end{array}
$$

and for $(q, p) = (2, 3)$ and $(q, p) = (3, 4)$, respectively, there exists exactly one A−stable diagonally implicit Runge–Kutta formula. These are given by

Print[" (q,p)=(2,3)"]

Grid$\left[\left\{\left\{\frac{1}{2} + \frac{1}{2\sqrt{3}}, 0, \frac{1}{2} + \frac{1}{2\sqrt{3}}\right\}, \left\{\frac{-1}{3}, \frac{1}{2} + \frac{1}{2\sqrt{3}}, \frac{1}{2} - \frac{1}{2\sqrt{3}}\right\}, \left\{\frac{1}{2}, \frac{1}{2}, \text{Null}\right\}\right\},\right.$
Dividers $\to$ {{False, False, True}, {False, False, True}}]
(q,p)=(2,3)

$$
\begin{array}{cc|cc}
\frac{1}{2}+\frac{1}{2\sqrt{3}} & 0 & \frac{1}{2}+\frac{1}{2\sqrt{3}} \\
-\frac{1}{3} & \frac{1}{2}+\frac{1}{2\sqrt{3}} & \frac{1}{2}-\frac{1}{2\sqrt{3}} \\
\hline
\frac{1}{2} & \frac{1}{2}
\end{array}
$$

and

```
Clear[α]
Print["(q,p) = (3,4),", "α = (2/√3)cos(π/18)"]
Grid[{{(1 + α)/2, 0, 0, (1 + α)/2}, {−α/2, (1 + α)/2, 0, 1/2},
{1 + α, −(1 + 2α), (1 + α)/2, (1 − α)/2},
{1/ (6α²) , 1 − 1/ (3α²) , 1/ (6α²) , Null}},
Dividers→{{False,False,False,True},{False,False,False,True}}
(q,p)=(3,4),   α = (2/√3)cos(π/18)
```

$$
\begin{array}{ccc|c}
\frac{1+\alpha}{2} & 0 & 0 & \frac{1+\alpha}{2} \\
-\frac{\alpha}{2} & \frac{1+\alpha}{2} & 0 & \frac{1}{2} \\
1+\alpha & -1-2\alpha & \frac{1+\alpha}{2} & \frac{1-\alpha}{2} \\
\hline
\frac{1}{6\alpha^2} & 1-\frac{1}{3\alpha^2} & \frac{1}{6\alpha^2}
\end{array}
$$

The aim of the present subsection is to introduce a semi-explicit Runge–Kutta method $(q, p) = (2, 4)$.

We use the following relations to pass from a step to the next:

$$
x_{n+1} = x_n + h_n \sum_{i=1}^{q} b_i k_{i,n}, \quad k_{i,n} = f\left(\arg_{i,n}\right),
$$

$$
\arg_{i,n} = x_n + h_n \left[\sum_{j=1}^{q} \lambda_{i,j} k_{j,n-1} + \sum_{j=1}^{q} a_{i,j} k_{j,n} \right], \tag{7.4}
$$

$$
k_{i,-1} = x_0.
$$

If we take

$$
\mu_i = \sum_{j=1}^{q} \left(\lambda_{i,j} + a_{i,j}\right),
$$

then instead of the first table, we get the next one

```
a=.;b=.;

Grid[{{λ_{1,1}, . . . , λ_{1,q}, a_{1,1}, . . . , a_{1,q}, μ_1}, {:, ,:,:, ,:,:},
{λ_{q,1}, . . . , λ_{q,q}, a_{q,1}, . . . , a_{q,q}, μ_q}, {, , , b_1, . . . , b_q, }},
Dividers→{{False,False,False,False,False,False,True},
{False,False,False,True}}]
```

$$\begin{array}{cccccc|c}
\lambda_{1,1} & \ldots & \lambda_{1,q} & a_{1,1} & \ldots & a_{1,q} & \mu_1 \\
\vdots & & \vdots & \vdots & & \vdots & \vdots \\
\lambda_{q,1} & \ldots & \lambda_{q,q} & a_{q,1} & \ldots & a_{q,q} & \mu_q \\
\hline
& & & b_1 & \ldots & b_q &
\end{array}$$

It is clear that if $\lambda_{i,j} = 0$, then (7.4) reduces (7.3).

To achieve the order 4 in 2 stages, we expand $x_{n+1} \approx x(t_n + h_n)$ in Taylor polynomials similar to [49]. To get the fourth order of accuracy, the next equalities have to be true.

$$\lambda_{1,1} + \lambda_{1,2} + a_{1,1} + a_{1,2} = \mu_1,$$

$$\lambda_{2,1} + \lambda_{2,2} + a_{2,1} + a_{2,2} = \mu_2,$$

$$b_1 + b_2 = 1, \tag{7.5}$$

$$b_1\mu_1 + b_2\mu_2 = 1/2, \tag{7.6}$$

$$b_1\mu_1^2 + b_2\mu_2^2 = 1/3, \tag{7.7}$$

$$b_1\mu_1^3 + b_2\mu_2^3 = 1/4, \tag{7.8}$$

$$b_1(a_{1,1}\mu_1 + a_{1,2}\mu_2) + b_2(a_{2,1}\mu_1 + a_{2,2}\mu_2) = 1/6, \tag{7.9}$$

$$b_1\mu_1(a_{1,1}\mu_1 + a_{1,2}\mu_2) + b_2\mu_2(a_{2,1}\mu_1 + a_{2,2}\mu_2) = 1/8, \tag{7.10}$$

$$b_1(a_{1,1}\mu_1^2 + a_{1,2}\mu_2^2) + b_2(a_{2,1}\mu_1^2 + a_{2,2}\mu_2^2) = 1/12,$$

$$(b_1a_{1,1} + b_2a_{2,1})(a_{1,1}\mu_1 + a_{1,2}\mu_2) +$$

$$(b_1a_{1,2} + b_2a_{2,2})(a_{2,1}\mu_1 + a_{2,2}\mu_2) = 1/24.$$

From (7.5), (7.6), (7.7), and (7.8) it results

$$b_1 = b_2 = 1/2, \quad \mu_1 = \left(3 - \sqrt{3}\right)\Big/3, \quad \mu_2 = \left(3 + \sqrt{3}\right)\Big/3,$$

and from (7.9)–(7.10) it follows

$$a_{1,1}\mu_1 + a_{1,2}\mu_2 = \left(2 - \sqrt{3}\right)\Big/12, \tag{7.11}$$

$$a_{2,1}\mu_1 + a_{2,2}\mu_2 = \left(2 + \sqrt{3}\right)\Big/12, \tag{7.12}$$

$$a_{1,1} + a_{2,1} = \mu_2, \tag{7.13}$$

$$a_{1,2} + a_{2,2} = \mu_1. \tag{7.14}$$

Then from (7.11), (7.12), (7.13), and (7.14) we get

$$a_{2,2} = \left(3 - \sqrt{3}\right)\Big/ 6 - \gamma,$$

$$a_{1,1} = \left(2 - \sqrt{3}\right)\Big/ 12 - \left(2 + \sqrt{3}\right)\gamma,$$

$$a_{2,1} = \left(1 + \sqrt{3}\right)\Big/ 4 + \left(2 + \sqrt{3}\right)\gamma,$$

$$a_{1,2} = \gamma, \quad \gamma \text{ real parameter.}$$

If we choose $\gamma = a_{1,2}$, then we get a semi-explicit two-step Runge–Kutta method of the fourth order, i.e., $(q,p) = (2,4)$. In that case we take $\lambda_{0,1} = \sqrt{3}\big/ 12$ and $\lambda_{1,1} = -\sqrt{3}\big/ 12$.

Chapter 8
Pi Formulas

Perhaps number π is the most famous and fascinating number in mathematics.

A long list of papers and books is dedicated to this number. We mention only some of them: [2, 3, 6–9, 17, 18, 30, 63, 69, 71, 76], and [77]. More specific references will be given as the context requires.

Along this chapter, we revise some results related to π and implicitly show how powerful is *Mathematica* when handling very complicated expressions.

π is intimately related to the properties of circles and spheres in elementary mathematics. At the advanced level, number π is present in complicated results not obviously tied to mathematics. For a planar circle of radius r, its circumference and area are given by

$$C = 2\pi r \ \text{ and } \ A = \pi r^2.$$

8.1 Various Simple and Not So Simple Formulas

8.1.1 Vandermonde Identity

vandermonde$=\sum_{k=0}^{\infty} \left(\frac{(2k-3)!!}{(2k)!!} \right)^2$

$\frac{4}{\pi}$

Regarding the speed of convergence we note some remarks.

© Springer International Publishing AG 2017
M. Mureşan, *Introduction to Mathematica® with Applications*,
DOI 10.1007/978-3-319-52003-2_8

Fig. 8.1 Vandermonde formula

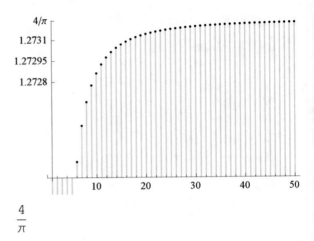

$$\text{vandermonde}[n\_]:=\sum_{k=0}^{n}\left(\frac{(2k-3)!!}{(2k)!!}\right)^2$$

Table[N[vandermonde[k]],{k,6}]   (* The first six terms *)
DiscretePlot[vandermonde[n],{n,50},
Ticks→{Automatic,{1.2728,1.27295,1.2731,{4/π,"4/π"}}},
PlotStyle→Black,ImageSize→250]
Simplify[vandermonde[∞]]
{1.25,1.26563,1.26953,1.27106,1.2718,1.27223}

See Fig. 8.1.

8.1.2 Sums Connected to Polygonal Numbers

A *polygonal number* is a number represented as dots or pebbles arranged in the shape of a regular polygon. Some series of this kind follow.

$$\{\text{spol1}=\sum_{k=1}^{\infty}\frac{3}{k(2k-1)(4k-3)},$$
$$\text{spol2}=\sum_{k=1}^{\infty}\frac{3\sqrt{3}}{(3k-1)(3k-2)},$$
$$\text{spol3}=4\sqrt{3}\sum_{k=1}^{\infty}\frac{12k-5}{8k(2k-1)(3k-1)(6k-5)},$$
$$\text{spol4}=16\sum_{k=1}^{\infty}\frac{864k(k-1)+226}{(12k-1)(12k-5)(12k-7)(12k-11)(4k-1)(4k-3)}\}$$
$$\{\pi,\pi,\pi,\pi\}$$

8.1.3 Machbin's and Machbin-Like Formulas

Relevant information on this topic may be found in many papers. We only mention [70] and [76].

```
FullSimplify[{4ArcTan[1/5] − ArcTan [1/239] ,    (* Machbin's original
formula 1706 *)
8ArcTan[1/10] − ArcTan [1/239] − 4ArcTan [1/515],    (* Klingenstierna
1730 *)
2ArcTan[1/2] − ArcTan [1/7], ArcTan[1/2] + ArcTan [1/3],    (* Euler 1738 *)
5ArcTan[1/7] + 2ArcTan [3/79],    (* Euler 1755 *)
4ArcTan[1/5] − ArcTan [1/70] + ArcTan [1/99],    (* Euler 1764 *)
ArcTan[1/2] + ArcTan [1/3],    (* Hutton 1776 *)
2ArcTan[1/3] + ArcTan [1/7],    (* Hutton 1776 *)
ArcTan[1/2] + ArcTan [1/5] + ArcTan [1/8],    (* Strassnitzky 1844 *)
12ArcTan[1/18] + 8ArcTan [1/57] − 5ArcTan [1/239],    (* Gauss 1863 *)
6ArcTan[1/8] + 2ArcTan [1/57] + ArcTan [1/239],    (* Störmer 1896 *)
5ArcTan [1/7] + 2ArcTan [3/79]}]
{π/4, π/4, π/4, π/4, π/4, π/4, π/4, π/4, π/4, π/4, π/4, π/4}
```

8.1.3.1   Kanada

In December 2002, Kanada computed π to over 1.24 trillion decimal digits. His team first computed π in hexadecimal (base 16) to 1,030,700,000,000 places, using the following two arctangent relations:

```
FullSimplify[{12ArcTan [1/49] + 32ArcTan [1/57] − 5ArcTan [1/239] +
12ArcTan [1/110443],    (* Takano, [63] *)
44ArcTan [1/57] + 7ArcTan [1/239] − 12ArcTan [1/682] + 24ArcTan [1/12943]}]
(* Störmer 1896, [12] *)
{π/4, π/4}
```

8.1.4   Gregory and Leibniz Formula

This formula may be found in [71, (32)]. The convergence of this series is slow. The graph below also shows the low speed of convergence.

```
Clear[s,n]
s[n_]:=4∑ⁿ_{k=0} (−1)ᵏ/(2k+1);
DiscretePlot[s[n],{n,50},
Ticks→{Automatic,{3,3.1,π,3.2,3.3}},
PlotStyle→Black,ImageSize→250]
s[∞]    (* The sum of the series *)
```

See Fig. 8.2.

Fig. 8.2 Gregory and
Leibniz formula

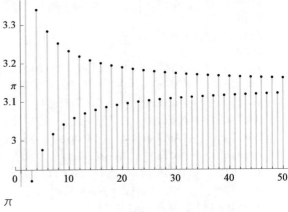

Fig. 8.3 Vardi formula

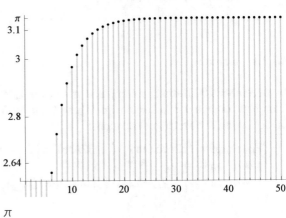

8.1.5   Vardi Formula

This formula may be found in [71, (8)]. In this formula number π is connected to
the ζ function of Riemann and is computed by it.

```
vardi[n_]:=∑ⁿₖ₌₁ (3ᵏ−1)/4ᵏ Zeta[k+1];
Table[N[vardi[k]],{k,1,26,5}]   (* Some terms of the sequence *)
DiscretePlot[vardi[n],{n,50},
Ticks→{Automatic,{2.64,2.8,3,3.1,π}},
PlotStyle→Black,ImageSize→250]
vardi[∞]   (* The sum of the series *)
{0.822467,2.60687,3.01488,3.11152,3.13446,3.1399}
```

See Fig. 8.3.

Fig. 8.4 Abraham Sharp
formula

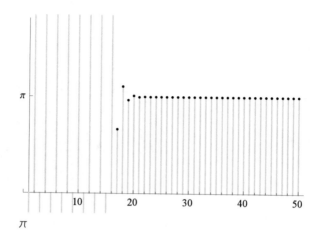

8.1.6 Abraham Sharp Formula

This formula may be found in [71, (9)].

```
s[n_]:=√12 ∑ⁿₖ₌₀ (−1)ᵏ/(3ᵏ(2k+1));
N[Table[s[n]-π,{n,10}]]   (* We note the speed of convergence
toward zero *)
DiscretePlot[s[n],{n,50},
Ticks→{Automatic,{{π,"π"}}},
PlotStyle→Black,ImageSize→250]
s[∞]   (* The sum of the series *)
{-0.0623912,0.0145888,-0.00373976,0.00101209,-0.000283868,
0.0000816591,-0.0000239376,7.1202217131194345*^-6,
-2.1426517125000544*^-6,6.509132886023394*^-7}
```

See Fig. 8.4.

8.1.7 Not So Simple Series

A good source of series connected to π is [71]. For some of them, we offer at
least a method of computation. In some cases, we use a second method based on an
appropriately chosen definite integral.

Fig. 8.5 Case 1

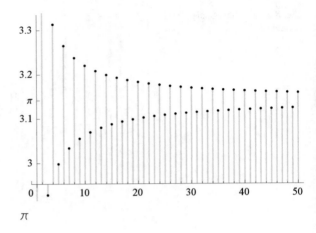

8.1.7.1 Case 1

case1[n_]:=$\frac{3\sqrt{3}}{2}\sum_{k=0}^{n}(-1)^k\left(\frac{1}{3k+1}+\frac{1}{3k+2}\right)$;
Table[N[case1[k]],{k,6}]   (* The first six terms of the sequence *)
DiscretePlot[case1[n],{n,50},
Ticks→{Automatic,{3,3.1,π,3.2,3.3}},
PlotStyle→Black,ImageSize→250]
case1[∞]   (* The sum of the series *)
{2.72798,3.42389,2.9279,3.31333,2.99812,3.26476}

See Fig. 8.5.

We can prove the previous result in the following way.

Clear[a,b]
$\sum_{k=0}^{\infty}(-1)^k\left(\frac{a}{3k+1}+\frac{b}{3k+2}\right)$
FullSimplify[%]
$\frac{1}{9}\left(\sqrt{3}a\pi+\sqrt{3}b\pi+a\text{Log}[8]-b\text{Log}[8]\right)$
$\frac{1}{9}\left(\sqrt{3}(a+b)\pi+(a-b)\text{Log}[8]\right)$

We obviously choose $a = b$ and have

$\sum_{k=0}^{\infty}(-1)^k\left(\frac{a}{3k+1}+\frac{a}{3k+2}\right)$
$\frac{2a\pi}{3\sqrt{3}}$

The result is clear.

8.1.7.2 Case 2

We treat the next series in a similar way.

```
Clear[a,b,c]
```
$\sum_{k=0}^{\infty}(-1)^k \left(\frac{a}{4k+1} + \frac{b}{4k+2} + \frac{c}{4k+3}\right)$
```
FullSimplify[%]
```
$\frac{1}{16}\left(a\pi + c\pi + 2b\pi\text{Cot}\left[\frac{\pi}{8}\right] + a\pi\text{Cot}\left[\frac{\pi}{8}\right]^2 + c\pi\text{Cot}\left[\frac{\pi}{8}\right]^2 + \right.$

$4\sqrt{2}a\text{Cot}\left[\frac{\pi}{8}\right]\text{Log}\left[\text{Cos}\left[\frac{\pi}{8}\right]\right] - 4\sqrt{2}c\text{Cot}\left[\frac{\pi}{8}\right]\text{Log}\left[\text{Cos}\left[\frac{\pi}{8}\right]\right] - $

$\left. 4\sqrt{2}a\text{Cot}\left[\frac{\pi}{8}\right]\text{Log}\left[\text{Sin}\left[\frac{\pi}{8}\right]\right] + 4\sqrt{2}c\text{Cot}\left[\frac{\pi}{8}\right]\text{Log}\left[\text{Sin}\left[\frac{\pi}{8}\right]\right]\right)\text{Tan}\left[\frac{\pi}{8}\right]$

$\frac{1}{8}\left(\left(\sqrt{2}a + b + \sqrt{2}c\right)\pi + 2\sqrt{2}(a - c)\text{ArcCoth}\left[\sqrt{2}\right]\right)$

We consider that $c \to a$ and solve the equation

```
FullSimplify[Solve[ 1/8 ( √2a + b + √2a ) ==1,b]
```
$\left\{\left\{b \to 8 - 2\sqrt{2}a\right\}\right\}$

We conclude that the series is now

```
FullSimplify[ ∑_{k=0}^{∞}(-1)^k ( a/(4k+1) + (8-2√2a)/(4k+2) + a/(4k+3) )]
```
π

The last result shows that for any complex a one gets π. Particularly, for $a = 2\sqrt{2}$, we have

```
Simplify[ 2√2 Simplify[ ∑_{k=0}^{∞}(-1)^k ( 1/(4k+1) + 1/(4k+3) )]]
```
π

We want to see the speed of convergence of this series.

```
case2[n_]:=2√2 ∑_{k=0}^{n}(-1)^k ( 1/(4k+1) + 1/(4k+3) );
Table[N[case2[n]],{n,6}]   (* The first six terms of the sequence *)
DiscretePlot[case2[n],{n,50},
Ticks→{Automatic,{3,3.1,π,3.2}},
PlotStyle→Black,ImageSize→250]
Simplify[case2[∞]]
{2.80149, 3.37289, 2.96676, 3.282, 3.02434, 3.24223}
```

See Fig. 8.6.

A different approach is by means of a definite integral.

We introduce a different approach to this series based on a definite integral.

```
f[i_]:=∫_0^1 (x^{i-1})/(1+x^4) dx;
as=Array[a,3];
product= 4/√2 as.Array[f,3];
asArray=FullSimplify[product]
```
$\frac{1}{4}(\sqrt{2}\pi a[2] + 2\pi a[3] + 2a[1](\pi + 2\text{ArcCoth}[\sqrt{2}]) + 2a[3]\text{Log}[3 - 2\sqrt{2}])$

Because this intermediate result is complicated, we try to find relations between the values of transcendental numbers so that the result is simplified. We have found two such relations and we substitute them. Then

Fig. 8.6 Case 2

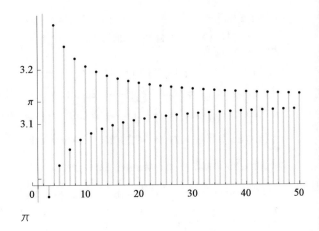

simpler=Simplify[Together[asArray/.{ArcCoth[$\sqrt{2}$]→(1/2)Log[3+2$\sqrt{2}$],
Log[3-2$\sqrt{2}$]→-Log[3+2$\sqrt{2}$]}]]
$\frac{1}{4}(\pi(2a[1] + \sqrt{2}a[2] + 2a[3]) + 2(a[1] - a[3])Log[3 + 2\sqrt{2}])$

GetTranscendentals[exp_]:=Union[Cases[exp,Pi|_ Log,Infinity]]
transsimpler=GetTranscendentals[simpler]
$\left\{\pi, \text{Log}\left[3 + 2\sqrt{2}\right]\right\}$

collected=Map[Factor,Collect[simpler,transsimpler]]
$\frac{1}{4}\pi\left(2a[1] + \sqrt{2}a[2] + 2a[3]\right) + \frac{1}{2}(a[1] - a[3])\text{Log}\left[3 + 2\sqrt{2}\right]$

We write the system of linear equations which follows.

syst=DeleteCases[Flatten[CoefficientList[collected,transsimpler]],0]
$\left\{\frac{a[1]}{2} - \frac{a[3]}{2}, \frac{a[1]}{2} + \frac{a[2]}{2\sqrt{2}} + \frac{a[3]}{2}\right\}$

The previous system is written under matrix form and we find its rank.

TableForm[syst,2]
(coeffMatrixx=Normal[CoefficientArrays[syst,as]][[2]])//MatrixForm
MatrixRank[coeffMatrixx]
$\frac{a[1]}{2} - \frac{a[3]}{2}$
$\frac{a[1]}{2} + \frac{a[2]}{2\sqrt{2}} + \frac{a[3]}{2}$
$\begin{pmatrix} \frac{1}{2} & 0 & -\frac{1}{2} \\ \frac{1}{2} & \frac{1}{2\sqrt{2}} & \frac{1}{2} \end{pmatrix}$
2

Since the rank is 2, one coefficient becomes parameter and the system is solved with respect to this parameter. Thus, we find out the next result.

```
sol=Solve[syst=={0,1}]
{{a[2] → 2√2 − 2√2a[1], a[3] → a[1]}}
```

We substitute and check the result.

$$\frac{4}{\sqrt{2}}\sum_{k=0}^{\infty}(-1)^k\left(\frac{a}{4k+1}+\frac{2\sqrt{2}-2\sqrt{2}a}{4k+2}+\frac{a}{4k+3}\right)$$

```
Simplify[%]
```

$$-\frac{\sqrt{2}(-1+\sqrt{2})\pi}{-2+\sqrt{2}}$$

The last series converges to π for any complex number a.

8.1.7.3 Case 3

Now we study the series

$$\sum_{k=0}^{\infty}(-1)^k\left(\frac{a}{5k+1}+\frac{b}{5k+2}+\frac{c}{5k+3}+\frac{d}{5k+4}\right);$$

```
FullSimplify[%]
```

$$\frac{1}{20\sqrt{5}}(2\sqrt{2}(\sqrt{5-\sqrt{5}}(b+c)+\sqrt{5+\sqrt{5}}(a+d))\pi+$$
$$10(a+b-c-d)\text{ArcCosh}\left[\tfrac{3}{2}\right]+\sqrt{5}(a-b+c-d)\text{Log}[16])$$

We solve the system of equations

```
Solve[a+b-c-d==0&&a-b+c-d==0,{c,d}]
{{c → b,d → a}}
```

and substitute

```
FullSimplify[%%%/.{c→b, d→a}]
```

$$\frac{1}{5}\sqrt{\frac{2}{5}}\left(\sqrt{5+\sqrt{5}}a+\sqrt{5-\sqrt{5}}b\right)\pi$$

If we consider $\{a, b, c, d\} = \{1, 0, 0, 1\}$, then

```
{a,b,c,d}={1,0,0,1};
```

$$\sum_{k=0}^{\infty}(-1)^k\left(\frac{a}{5k+1}+\frac{b}{5k+2}+\frac{c}{5k+3}+\frac{d}{5k+4}\right);$$

```
FullSimplify[%]
```

$$\frac{1}{5}\sqrt{2+\frac{2}{\sqrt{5}}}\,\pi$$

We note that for each complex a and b it is true that

```
Clear[a,b]
FullSimplify[∑_{k=0}^{∞}(-1)^k (a/5k+1 + b/5k+2 + b/5k+3 + a/5k+4)]
```

$$\frac{1}{5}\sqrt{\frac{2}{5}}\left(\sqrt{5+\sqrt{5}}a+\sqrt{5-\sqrt{5}}b\right)\pi$$

By a definite integral

We can try to approach this series by means of the corresponding definite integral.
We have

```
Clear[a]
```
$$f[i\_]:=\int_0^1 \frac{x^{i-1}}{1+x^5}\ dx;$$
```
as=Array[a,4];
as.Array[f,4];
asArray=FullSimplify[%]
```
$$\frac{1}{100}\left(2\sqrt{10}\pi\left(\sqrt{5-\sqrt{5}}(a[2]+a[3])+\sqrt{5+\sqrt{5}}(a[1]+a[4])\right)+\right.$$
$$\sqrt{5}a[2]\mathrm{ArcTanh}\left[\frac{6765\sqrt{5}}{15127}\right]+5a[1]\left(2\sqrt{5}\mathrm{ArcCoth}\left[\frac{3}{\sqrt{5}}\right]+\mathrm{Log}[16]\right)-$$
$$5\left(2\sqrt{5}a[4]\mathrm{ArcCoth}\left[\frac{3}{\sqrt{5}}\right]+a[3]\left(2\sqrt{5}\mathrm{ArcCoth}\left[\frac{3}{\sqrt{5}}\right]-\mathrm{Log}[16]\right)+\right.$$
$$\left.(a[2]+a[4])\mathrm{Log}[16])\right)$$

```
LogExpand[exp_]:=PowerExpand[exp]/.Log[n_Integer]:→
Apply[#2.Log[#1]&,
Transpose[FactorInteger[n]]]
GetTranscendentals[exp_]:=Union[Cases[exp,Pi| _ArcCoth| _ArcTanh|
_Log,Infinity]]
simpler=Together[
```
$$\mathrm{LogExpand}[asArray]/.\mathrm{ArcTanh}\left[\frac{6765\sqrt{5}}{15127}\right]\to 10\mathrm{ArcCoth}\left[\frac{3}{\sqrt{5}}\right]]$$
$$\frac{1}{50}\left(\sqrt{10\left(5+\sqrt{5}\right)}\pi a[1]+\sqrt{10\left(5-\sqrt{5}\right)}\pi a[2]+\sqrt{10\left(5-\sqrt{5}\right)}\pi a[3]+\right.$$
$$\sqrt{10\left(5+\sqrt{5}\right)}\pi a[4]+5\sqrt{5}a[1]\mathrm{ArcCoth}\left[\frac{3}{\sqrt{5}}\right]+5\sqrt{5}a[2]\mathrm{ArcCoth}\left[\frac{3}{\sqrt{5}}\right]-$$
$$5\sqrt{5}a[3]\mathrm{ArcCoth}\left[\frac{3}{\sqrt{5}}\right]-5\sqrt{5}a[4]\mathrm{ArcCoth}\left[\frac{3}{\sqrt{5}}\right]+$$
$$10a[1]\mathrm{Log}[2]-10a[2]\mathrm{Log}[2]+10a[3]\mathrm{Log}[2]-10a[4]\mathrm{Log}[2])$$

```
transsimpler=GetTranscendentals[simpler]
```
$$\left\{\pi,\mathrm{ArcCoth}\left[\frac{3}{\sqrt{5}}\right],\mathrm{Log}[2]\right\}$$

```
collected=Map[Factor,Collect[simpler,transsimpler]]
```
$$\frac{1}{5\sqrt{10}}\pi\left(\sqrt{5+\sqrt{5}}a[1]+\sqrt{5-\sqrt{5}}a[2]+\sqrt{5-\sqrt{5}}a[3]+\sqrt{5+\sqrt{5}}a[4]\right)+$$
$$\frac{(a[1]+a[2]-a[3]-a[4])\mathrm{ArcCoth}\left[\frac{3}{\sqrt{5}}\right]}{2\sqrt{5}}+\frac{1}{5}(a[1]-a[2]+a[3]-a[4])\mathrm{Log}[2]$$

```
system=DeleteCases[Flatten[CoefficientList[collected,
Append[transsimpler,Sqrt[2]]]],0]
TableForm[system,4]
(coeffMatrix=Normal[CoefficientArrays[system,as]][[2]])//MatrixForm
MatrixRank[coeffMatrix]
```

$$\left\{ \frac{a[1]}{5} - \frac{a[2]}{5} + \frac{a[3]}{5} - \frac{a[4]}{5}, \frac{a[1]}{2\sqrt{5}} + \frac{a[2]}{2\sqrt{5}} - \frac{a[3]}{2\sqrt{5}} - \frac{a[4]}{2\sqrt{5}}, \frac{1}{10}\sqrt{\frac{1}{5}\left(5+\sqrt{5}\right)}a[1] + \right.$$

$$\frac{1}{10}\sqrt{\frac{1}{5}\left(5-\sqrt{5}\right)}a[2] + \frac{1}{10}\sqrt{\frac{1}{5}\left(5-\sqrt{5}\right)}a[3] + \frac{1}{10}\sqrt{\frac{1}{5}\left(5+\sqrt{5}\right)}a[4]\right\}$$

$$\frac{a[1]}{5} - \frac{a[2]}{5} + \frac{a[3]}{5} - \frac{a[4]}{5}$$

$$\frac{a[1]}{2\sqrt{5}} + \frac{a[2]}{2\sqrt{5}} - \frac{a[3]}{2\sqrt{5}} - \frac{a[4]}{2\sqrt{5}}$$

$$\frac{1}{10}\sqrt{\frac{1}{5}\left(5+\sqrt{5}\right)}a[1] + \frac{1}{10}\sqrt{\frac{1}{5}\left(5-\sqrt{5}\right)}a[2] + \frac{1}{10}\sqrt{\frac{1}{5}\left(5-\sqrt{5}\right)}a[3]+$$

$$\frac{1}{10}\sqrt{\frac{1}{5}\left(5+\sqrt{5}\right)}a[4]$$

$$\begin{pmatrix} \frac{1}{5} & -\frac{1}{5} & \frac{1}{5} & -\frac{1}{5} \\ \frac{1}{2\sqrt{5}} & \frac{1}{2\sqrt{5}} & -\frac{1}{2\sqrt{5}} & -\frac{1}{2\sqrt{5}} \\ \frac{1}{10}\sqrt{\frac{1}{5}\left(5+\sqrt{5}\right)} & \frac{1}{10}\sqrt{\frac{1}{5}\left(5-\sqrt{5}\right)} & \frac{1}{10}\sqrt{\frac{1}{5}\left(5-\sqrt{5}\right)} & \frac{1}{10}\sqrt{\frac{1}{5}\left(5+\sqrt{5}\right)} \end{pmatrix}$$

3

```
sol=FullSimplify[Solve[system=={0,0,1}]]
```

$$\{\{a[2] \rightarrow \tfrac{1}{2}(5\sqrt{5+\sqrt{5}} - \left(1+\sqrt{5}\right)a[1]),$$

$$a[3] \rightarrow \tfrac{1}{2}\left(5\sqrt{5+\sqrt{5}} - \left(1+\sqrt{5}\right)a[1]\right), a[4] \rightarrow a[1]\}\}$$

Taking $\{a, b, c, d\} = \{1, 0, 0, 1\}$, we get the same result as before.

8.1.7.4 Case 4

Let us consider the next case.

```
Clear[a,b,c,d,e]
```

$$\sum_{k=0}^{\infty}(-1)^k \left(\frac{a}{6k+1} + \frac{b}{6k+2} + \frac{c}{6k+3} + \frac{d}{6k+4} + \frac{e}{6k+5}\right);$$

```
FullSimplify[%]
```

$$\tfrac{1}{36}((6a + 2\sqrt{3}b + 3c + 2\sqrt{3}d + 6e)\pi + 2(6\sqrt{3}(a-e)\mathrm{ArcCoth}[\sqrt{3}] + (b-d)\mathrm{Log}[8]))$$

Clearly there are two obvious options $e = a$ and $d = b$. Then

```
FullSimplify[%/.{e→a,d→b}]
```

$$\tfrac{1}{36}\left(12a + 4\sqrt{3}b + 3c\right)\pi$$

An almost easy option goes to $b = c = 0$ and $a = 1$, [71, (33)]. Then

$$\sum_{k=0}^{\infty}(-1)^k \left(\frac{1}{6k+1} + \frac{1}{6k+5}\right)$$

$$\frac{\pi}{3}$$

Fig. 8.7 Case 4

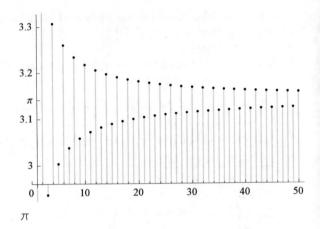

Another particular case is the next one, which is true for every complex c.

```
Clear[c]
```
$$\sum_{k=0}^{\infty}(-1)^k \left(\frac{1}{6k+1} + \frac{0}{6k+2} + \frac{c}{6k+3} + \frac{0}{6k+4} + \frac{1}{6k+5} \right)$$
$$\frac{1}{12}(4+c)\pi$$

Another easy option is $b=0$ and $4a+c=12$. Particularly, $a=1$ and $c=8$,

$$\sum_{k=0}^{\infty}(-1)^k \left(\frac{1}{6k+1} + \frac{8}{6k+3} + \frac{1}{6k+5} \right)$$
$$\pi$$

Regarding the convergence of this series, we remark the following:

```
case4[n_]:=∑ⁿₖ₌₀(-1)^k (1/(6k+1) + 8/(6k+3) + 1/(6k+5));
Table[N[case4[k]],{k,6}]   (* The first six terms of the sequence *)
DiscretePlot[case4[n],{n,50},
Ticks→{Automatic,{3,3.1,π,3.2,3.3}},
PlotStyle→Black,ImageSize→250]
Simplify[case4[∞]]
{2.74401, 3.41309, 2.93603, 3.30681, 3.00355, 3.2601}
```

See Fig. 8.7.

8.1.7.5　Case 5

Let us take a look at the series

```
Clear[a,b,c,d,e,f]
```
$$\sum_{k=0}^{\infty}(-1)^k \left(\frac{a}{7k+1} + \frac{b}{7k+2} + \frac{c}{7k+3} + \frac{d}{7k+4} + \frac{e}{7k+5} + \frac{f}{7k+6} \right);$$
```
FullSimplify[%]
```

$\frac{1}{28}\left(2f\pi\operatorname{Csc}\left[\frac{\pi}{7}\right] - b\operatorname{Log}[4] + c\operatorname{Log}[4] - d\operatorname{Log}[4] + e\operatorname{Log}[4] - f\operatorname{Log}[4] + \right.$
$4d\operatorname{Cos}\left[\frac{\pi}{7}\right]\operatorname{Log}\left[\operatorname{Cos}\left[\frac{3\pi}{14}\right]\operatorname{Cot}\left[\frac{\pi}{7}\right]\right] + 4(-b + e)\operatorname{Cos}\left[\frac{\pi}{7}\right] *$
$\operatorname{Log}\left[\operatorname{Root}\left[-8 + 20\#1 - 12\#1^2 + \#1^3\&, 2\right]\right] - 4f\operatorname{Cos}\left[\frac{\pi}{7}\right] *$
$\operatorname{Log}\left[\operatorname{Root}\left[-8 + 20\#1 - 12\#1^2 + \#1^3\&, 3\right]\right] + 4c\operatorname{Cos}\left[\frac{\pi}{7}\right] *$
$\operatorname{Log}\left[\operatorname{Sec}\left[\frac{3\pi}{14}\right]\operatorname{Tan}\left[\frac{\pi}{7}\right]\right] + 2c\pi\operatorname{Sec}\left[\frac{\pi}{14}\right] + 2d\pi\operatorname{Sec}\left[\frac{\pi}{14}\right] + 2b\pi\operatorname{Sec}\left[\frac{3\pi}{14}\right] +$
$2e\pi\operatorname{Sec}\left[\frac{3\pi}{14}\right] - 4f\operatorname{Log}\left[\operatorname{Root}\left[-8 + 20\#1 - 12\#1^2 + \#1^3\&, 2\right]\right]\operatorname{Sin}\left[\frac{\pi}{14}\right] +$
$4c\operatorname{Log}\left[\operatorname{Root}\left[-8 + 20\#1 - 12\#1^2 + \#1^3\&, 3\right]\right]\operatorname{Sin}\left[\frac{\pi}{14}\right] -$
$d\operatorname{Log}\left[\operatorname{Root}\left[-8 + 20\#1 - 12\#1^2 + \#1^3\&, 3\right]\right]\operatorname{Sin}\left[\frac{\pi}{14}\right] +$
$4b\operatorname{Log}\left[\operatorname{Root}\left[-1 + 12\#1 - 20\#1^2 + 8\#1^3\&, 3\right]\right]\operatorname{Sin}\left[\frac{\pi}{14}\right] -$
$4e\operatorname{Log}\left[\operatorname{Root}\left[-1 + 12\#1 - 20\#1^2 + 8\#1^3\&, 3\right]\right]\operatorname{Sin}\left[\frac{\pi}{14}\right] +$
$4\left(b\operatorname{Log}\left[\frac{1}{2}\operatorname{Csc}\left[\frac{\pi}{14}\right]^2\right] + d\operatorname{Log}\left[\frac{1}{2}\operatorname{Csc}\left[\frac{3\pi}{14}\right]^2\right] + e\operatorname{Log}\left[2\operatorname{Sin}\left[\frac{\pi}{14}\right]^2\right] +\right.$
$\left. c\operatorname{Log}\left[2\operatorname{Sin}\left[\frac{3\pi}{14}\right]^2\right] + f\operatorname{Log}\left[\operatorname{Sec}\left[\frac{3\pi}{14}\right]\operatorname{Tan}\left[\frac{\pi}{7}\right]\right]\right)\operatorname{Sin}\left[\frac{3\pi}{14}\right] +$
$a\left(2\pi\operatorname{Csc}\left[\frac{\pi}{7}\right] + \operatorname{Log}[4] - \operatorname{Cos}\left[\frac{\pi}{7}\right]\operatorname{Log}\left[16\operatorname{Sin}\left[\frac{\pi}{14}\right]^8\right] +\right.$
$4\operatorname{Log}\left[\operatorname{Root}\left[-8 + 20\#1 - 12\#1^2 + \#1^3\&, 2\right]\right]\operatorname{Sin}\left[\frac{\pi}{14}\right] +$
$\left.4\operatorname{Log}\left[\operatorname{Cos}\left[\frac{3\pi}{14}\right]\operatorname{Cot}\left[\frac{\pi}{7}\right]\right]\operatorname{Sin}\left[\frac{3\pi}{14}\right]\right)\right)$

We note that the result is rather complicated and it is hard to believe that it can be easily manipulated. Therefore, we pass to a particular form of it.

$7\sum_{k=0}^{\infty}(-1)^k\left(\frac{a}{7k+1} + \frac{b}{7k+2} + \frac{c}{7k+3} + \frac{c}{7k+4} + \frac{b}{7k+5} + \frac{a}{7k+6}\right);$
FullSimplify[%]
$\pi\left(a\operatorname{Csc}\left[\frac{\pi}{7}\right] + c\operatorname{Sec}\left[\frac{\pi}{14}\right] + b\operatorname{Sec}\left[\frac{3\pi}{14}\right]\right)$

FullSimplify$\left[\%/.\operatorname{Csc}\left[\frac{\pi}{7}\right] \to \operatorname{Sec}\left[\frac{\pi}{14}\right] + \operatorname{Sec}\left[\frac{3\pi}{14}\right]\right]$
$(a + c)\pi\operatorname{Sec}\left[\frac{\pi}{14}\right] + (a + b)\pi\operatorname{Sec}\left[\frac{3\pi}{14}\right]$

So the result

$\pi\left((a + c)\operatorname{Sec}\left[\frac{\pi}{14}\right] + (a + b)\operatorname{Sec}\left[\frac{3\pi}{14}\right]\right)$

depends on the (complex) coefficients a, b, and c.

8.1.7.6 Case 6

Let us consider the series

Clear[a,b,c,d]
$\sum_{k=0}^{\infty}(-1)^k\left(\frac{a}{8k+1} + \frac{b}{8k+2} + \frac{c}{8k+3} + \frac{d}{8k+4} + \frac{c}{8k+5} + \frac{b}{8k+6} + \frac{a}{8k+7}\right);$
sum8=FullSimplify[%]
$\frac{1}{16}\pi\left(2\sqrt{2}b + d + 2(a + (-1 + \sqrt{2})c)\operatorname{Csc}\left[\frac{\pi}{8}\right]\right)$

FullSimplify$\left[\text{sum8}/.\operatorname{Csc}\left[\frac{\pi}{8}\right] \to \sqrt{2}\sqrt{2 + \sqrt{2}}\right]$
$\frac{1}{16}\left(2\sqrt{2}b + 2\sqrt{2\left(2 + \sqrt{2}\right)}\left(a + \left(-1 + \sqrt{2}\right)c\right) + d\right)\pi$

Then for $a = b = c = 0$, and $d = 1$, we get

```
sum8/.{a→0,b→0,c→0,d→1}
```
$$\frac{\pi}{16}$$

This is precisely the Gregory and Leibniz formula.

If we consider the above series under its general form, we have

$$\sum_{k=0}^{\infty}(-1)^k\left(\frac{a}{8k+1} + \frac{b}{8k+2} + \frac{c}{8k+3} + \frac{d}{8k+4} + \frac{e}{8k+5} + \frac{f}{8k+6} + \frac{g}{8k+7}\right);$$

```
FullSimplify[Together[%/.{Cos[π/8] → √(2+√2)/2,
```

$$\text{Sin}\left[\frac{\pi}{8}\right] \to \frac{\sqrt{2-\sqrt{2}}}{2}, \text{Tan}\left[\frac{\pi}{8}\right] \to \frac{\sqrt{2-\sqrt{2}}}{\sqrt{2+\sqrt{2}}}, \text{Cot}\left[\frac{\pi}{8}\right] \to \frac{\sqrt{2+\sqrt{2}}}{\sqrt{2-\sqrt{2}}},$$

$$\text{Csc}\left[\frac{\pi}{8}\right] \to \sqrt{2}\sqrt{2+\sqrt{2}}, \text{Tan}\left[\frac{\pi}{16}\right] \to -1 - \sqrt{2} + \sqrt{2\left(2+\sqrt{2}\right)},$$

$$\text{Cot}\left[\frac{\pi}{16}\right] \to 1 + \sqrt{2} + \sqrt{2\left(2+\sqrt{2}\right)}, \text{Tan}\left[\frac{3\pi}{16}\right] \to 1 - \sqrt{2} + \sqrt{2\left(2-\sqrt{2}\right)},$$

$$\text{Cot}\left[\frac{3\pi}{16}\right] \to -1 + \sqrt{2} + \sqrt{2\left(2-\sqrt{2}\right)}\}]];$$

The result is rather complicated and is out of our goal.

By a definite integral

Because the last result is rather long, we approach this series by means of a definite integral.

```
Clear[a]
f[i_]:=∫₀¹ x^(i-1)/(1+x^8) dx;
as=Array[a,7];
asArray=FullSimplify[as.Array[f,7]]
```

$$\frac{1}{16}\left(\pi\left(\sqrt{2}a[2] + a[4] + \sqrt{4-2\sqrt{2}}(a[3]+a[5]) + \sqrt{2}a[6] + \right.\right.$$
$$\left.\left.\sqrt{2(2+\sqrt{2})}(a[1]+a[7])\right)\right) -$$
$$\sqrt{2}a[6]\text{ArcCosh}[3] + 4\left((a[1]-a[7])\text{ArcTanh}\left[\text{Cos}\left[\frac{\pi}{8}\right]\right] + \right.$$
$$\left.(-a[3]+a[5])\text{ArcTanh}\left[\text{Sin}\left[\frac{\pi}{8}\right]\right]\right)\text{Cos}\left[\frac{\pi}{8}\right] -$$
$$\sqrt{2}a[2]\text{Log}\left[3-2\sqrt{2}\right] + 4\left((a[3]-a[5])\text{ArcTanh}\left[\text{Cos}\left[\frac{\pi}{8}\right]\right] + \right.$$
$$\left.(a[1]-a[7])\text{ArcTanh}\left[\text{Sin}\left[\frac{\pi}{8}\right]\right]\right)\text{Sin}\left[\frac{\pi}{8}\right]$$

```
LogExpand[exp_]:=PowerExpand[exp]/.Log[n_Integer]:→
Apply[#2.Log[#1]&,Transpose[FactorInteger[n]]]
GetTranscendentals[exp_]:=
Union[Cases[exp,Pi |_ArcTanh |_ArcCosh |_Log |_Cos |_Sin, Infinity]]
simpler=
```
$$Together\left[LogExpand[asArray]/.\left\{ArcCosh[3] \to Log\left[3+2\sqrt{2}\right],\right.\right.$$
$$\left.\left.Log\left[3-2\sqrt{2}\right] \to -Log\left[3+2\sqrt{2}\right]\right\}\right];$$

```
transsimpler=GetTranscendentals[simpler]
```
$$\left\{\pi, ArcTanh\left[Cos\left[\tfrac{\pi}{8}\right]\right], ArcTanh\left[Sin\left[\tfrac{\pi}{8}\right]\right], Cos\left[\tfrac{\pi}{8}\right], Log\left[3+2\sqrt{2}\right],\right.$$
$$\left.Sin\left[\tfrac{\pi}{8}\right]\right\}$$

```
collected=Map[Factor,Collect[simpler,transsimpler]];
system=DeleteCases[Flatten[CoefficientList[collected,
Append[transsimpler,Sqrt[2]]]],0]
TableForm[system,7];
(coeffMatrix=Normal[CoefficientArrays[system,as]][[2]])//MatrixForm
MatrixRank[coeffMatrix]
```

$$\begin{pmatrix}
0 & \frac{1}{16} & 0 & 0 & 0 & -\frac{1}{16} & 0 \\
\frac{1}{4} & 0 & 0 & 0 & 0 & 0 & -\frac{1}{4} \\
0 & 0 & -\frac{1}{4} & 0 & \frac{1}{4} & 0 & 0 \\
0 & 0 & \frac{1}{4} & 0 & -\frac{1}{4} & 0 & 0 \\
\frac{1}{4} & 0 & 0 & 0 & 0 & 0 & -\frac{1}{4} \\
0 & 0 & 0 & \frac{1}{16} & 0 & 0 & 0 \\
\frac{\sqrt{2+\sqrt{2}}}{16} & \frac{1}{16} & \frac{\sqrt{2-\sqrt{2}}}{16} & 0 & \frac{\sqrt{2-\sqrt{2}}}{16} & \frac{1}{16} & \frac{\sqrt{2+\sqrt{2}}}{16}
\end{pmatrix}$$

5

We consider a particular case.

```
sol=Solve[system=={0,0,0,0,0,1,0}]
```
$$\{\{a[2] \to -\sqrt{2+\sqrt{2}}\,a[1] - \sqrt{2-\sqrt{2}}\,a[3], a[4] \to 16,$$
$$a[5] \to a[3], a[6] \to -\sqrt{2+\sqrt{2}}\,a[1] - \sqrt{2-\sqrt{2}}\,a[3], a[7] \to a[1]\}\}$$

If $a[1] = a[3] = 0$, then

```
sol/.{a[1]→0,a[3]→0}
```
$$\{\{a[2] \to 0, a[4] \to 16, a[5] \to 0, a[6] \to 0, a[7] \to 0\}\}$$

i.e.,

$$16 \sum_{k=0}^{\infty} \frac{(-1)^k}{8k+4}$$
$$\pi$$

The last series is the Gregory and Leibniz series.

8.1.7.7   Case 7

Now we consider the following series:

```
Clear[a,b,c,d,e,f,g,h]
```
$$\sum_{k=0}^{\infty}(-1)^k \left(\frac{a}{9k+1} + \frac{b}{9k+2} + \frac{c}{9k+3} + \frac{d}{9k+4} + \frac{e}{9k+5} + \frac{f}{9k+6} + \frac{g}{9k+7} + \frac{h}{9k+8}\right);$$
```
FullSimplify[%];
```
$$\sum_{k=0}^{\infty}(-1)^k \left(\frac{a}{9k+1} + \frac{b}{9k+2} + \frac{c}{9k+3} + \frac{d}{9k+4} + \frac{d}{9k+5} + \frac{c}{9k+6} + \frac{b}{9k+7} + \frac{a}{9k+8}\right);$$
```
FullSimplify[%]
```
$$\frac{1}{27}\pi \left(2\sqrt{3}c + 3\left(a\,\mathrm{Csc}\left[\frac{\pi}{9}\right] + b\,\mathrm{Csc}\left[\frac{2\pi}{9}\right] + d\,\mathrm{Sec}\left[\frac{\pi}{18}\right]\right)\right)$$

An easy choice for the second series is $a = b = d = 0$. Then one has

$$\sum_{k=0}^{\infty}(-1)^k \left(\frac{1}{9k+3} + \frac{1}{9k+6}\right)$$
$$\frac{2\pi}{9\sqrt{3}}$$

The last series coincides up to a factor with the series in Case 1.

8.1.7.8   Case 8

By [71, (34)] one has

$$\sum_{k=0}^{\infty}(-1)^k \left(\frac{1}{10k+1} + \frac{-1}{10k+3} + \frac{1}{10k+5} + \frac{-1}{10k+7} + \frac{1}{10k+9}\right)$$
$$\frac{\pi}{4}$$

We want to take a closer look at the series

```
Clear[a,b,c,d,e]
```
$$\sum_{k=0}^{\infty}(-1)^k \left(\frac{a}{10k+1} + \frac{b}{10k+2} + \frac{c}{10k+3} + \frac{d}{10k+4} + \frac{e}{10k+5} + \frac{d}{10k+6} + \right.$$
$$\left.\frac{c}{10k+7} + \frac{b}{10k+8} + \frac{a}{10k+9}\right);$$

```
FullSimplify[%]
```
$$\frac{1}{100}\left(10\left(1 + \sqrt{5}\right)a + 2\sqrt{10\left(5 + \sqrt{5}\right)}b + 10\left(-1 + \sqrt{5}\right)c + \right.$$
$$\left.2\sqrt{50 - 10\sqrt{5}}d + 5e\right)\pi$$

We may choose the following particular case

```
FullSimplify[%/.{a→1,b→0,c→-1,d→0,e→0}]
```
$$\frac{\pi}{5}$$

Regarding the convergence of the initial series, we note that

```
case10[n_]:=∑_{k=0}^{∞}(-1)^k ( 1/(10k+1) + -1/(10k+3) + 1/(10k+5) + -1/(10k+7) + 1/(10k+9))
Table[N[case10[k]],{k,6}]   (* The first six terms *)
DiscretePlot[case10[n],{n,50},
Ticks→{Automatic,{.78,{π/4,"π/4"},.79}},
PlotStyle→Black,ImageSize→250]
```

Fig. 8.8 Case 8

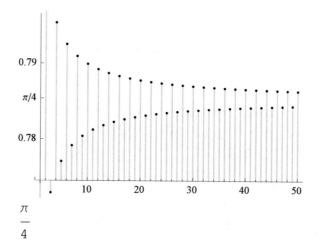

$$\frac{\pi}{4}$$

Simplify[case10[∞]]   (* The sum of the series *)
{0.76046, 0.802046, 0.772906, 0.795394, 0.777067, 0.79254}

See Fig. 8.8.

8.1.7.9 Case 9

Let us consider the series

Clear[a,b,c,d,e]
$$\sum_{k=0}^{\infty}(-1)^k \left(\frac{a}{11k+1} + \frac{b}{11k+2} + \frac{c}{11k+3} + \frac{d}{11k+4} + \frac{e}{11k+5} + \frac{e}{11k+6} + \frac{d}{11k+7} + \frac{c}{11k+8} + \frac{b}{11k+9} + \frac{a}{11k+10}\right);$$

sum11=FullSimplify[%]
$$\frac{1}{11}\pi \left(a \operatorname{Csc}\left[\tfrac{\pi}{11}\right] + b \operatorname{Csc}\left[\tfrac{2\pi}{11}\right] + e \operatorname{Sec}\left[\tfrac{\pi}{22}\right] + d \operatorname{Sec}\left[\tfrac{3\pi}{22}\right] + c \operatorname{Sec}\left[\tfrac{5\pi}{22}\right]\right)$$

Now we choose a particular case

FullSimplify[%/.{b → 0, c → 0, d → 0, e → 0}]
$$\frac{1}{11}a\pi \operatorname{Csc}\left[\tfrac{\pi}{11}\right]$$

It means that

$$\sum_{k=0}^{\infty}(-1)^k \left(\frac{1}{11k+1} + \frac{1}{11k+10}\right)$$ (* Equals *)
$$\frac{1}{11}\pi \operatorname{Csc}\left[\tfrac{\pi}{11}\right]$$

Also, one has

FullSimplify[sum11/.{a→1,b→0,c→1,d→0,e→-1,f→0}]
$$\frac{\pi}{2\sqrt{2}}$$

8.1.7.10   Case 10

Because the amount of computation is huge for the next series, we consider a simplified version of it, namely,

Clear[a,b,c,d,e,f]

$$\sum_{k=0}^{\infty}(-1)^k\left(\frac{a}{12k+1}+\frac{b}{12k+2}+\frac{c}{12k+3}+\frac{d}{12k+4}+\frac{e}{12k+5}+\frac{f}{12k+6}+\frac{e}{12k+7}+\frac{d}{12k+8}+\frac{c}{12k+9}+\frac{b}{12k+10}+\frac{a}{12k+11}\right);$$

FullSimplify[%]

$$\frac{1}{72}(6(\sqrt{2}+\sqrt{6})a+12b+6\sqrt{2}c+4\sqrt{3}d+6\sqrt{2}(-1+\sqrt{3})e+3f)\pi$$

We have a simple series

FullSimplify[%/.{a→1,e→1,b→-1,c→-1,d→ $\sqrt{3/2}$,f→4}]

$\frac{\pi}{\sqrt{6}}$

By definite integral

The general case is discussed using a definite integral.

```
f[i_]:=∫₀¹ x^{i-1}/(1+x^12) dx;
as=Array[a,11];
as.Array[f,11];
asArray=FullSimplify[TrigToExp[%]];
LogExpand[exp_]:=PowerExpand[exp]/.Log[n_Integer]:→
Apply[#2.Log[#1]&,Transpose[FactorInteger[n]]]
GetTranscendentals[exp_]:=
Union[Cases[exp,Pi | _ArcCosh | _Log | _ ArcTanh,Infinity]]
simpler=Together[LogExpand[asArray]/.{ArcCosh[19601]→
ArcCosh[17]-Log[577-408√2],ArcCosh[577]→-Log[577-408√2],
Log[17-12√2]→Log[3-2√2]+Log[2], ArcCosh[49]→-Log[49-20√6],
ArcCosh[3]→-Log[3-2√2],ArcTanh[2√6/5] →ArcCosh[5]}];

transsimpler=GetTranscendentals[simpler];
collected=Map[Factor,Collect[simpler,transsimpler]];
system=DeleteCases[Flatten[CoefficientList[collected,
Append[transsimpler,Sqrt[2]]]],0]
TableForm[system,11]
(coeffMatrix=Normal[CoefficientArrays[system,as]][[2]])//MatrixForm
MatrixRank[coeffMatrix]
system/.{a[1]→1,a[2]→0,a[3]→1,a[4]→0,a[5]→-1,a[6]→0,
a[7]→-1,a[8]→0,a[9]→1,a[10]→0,a[11]→1}
```

Solve
$$\left[\text{system} == \left\{ 0, -\frac{1}{16\sqrt{3}}, 0, -\frac{1}{72}, 0, 0, -\frac{1}{36}, 0, -\frac{1}{8\sqrt{3}}, 0, \frac{\sqrt{2+\sqrt{3}}}{6}, \frac{1}{6} - \frac{1}{4\sqrt{3}} \right\} \right]$$
$\{\{a[1] \to 1, a[2] \to 0, a[3] \to 1, a[5] \to -1, a[6] \to -\frac{4a[4]}{\sqrt{3}}, a[7] \to -1,$
$a[8] \to a[4], a[9] \to 1, a[10] \to 0, a[11] \to 1\}\}$

We check the result

asa={1,0,1,0,-1,0,-1,0,1,0,1};
asa.Array[f,11];
2Sqrt[2]FullSimplify[TrigToExp[%]]
π

8.1.7.11   Case 11

We will study the next series

Clear[a,b,c,d,e,f]
$\sum_{k=0}^{\infty}(-1)^k \left(\frac{a}{13k+1} + \frac{b}{13k+2} + \frac{c}{13k+3} + \frac{d}{13k+4} + \frac{e}{13k+5} + \frac{f}{13k+6} + \frac{f}{13k+7} + \frac{e}{13k+8} + \frac{d}{13k+9} + \frac{c}{13k+10} + \frac{b}{13k+11} + \frac{a}{13k+12} \right);$

FullSimplify[%]

$\frac{1}{13}\pi \left(a \, \text{Csc}\left[\frac{\pi}{13}\right] + b \, \text{Csc}\left[\frac{2\pi}{13}\right] + c \, \text{Csc}\left[\frac{3\pi}{13}\right] + f \, \text{Sec}\left[\frac{\pi}{26}\right] + e \, \text{Sec}\left[\frac{3\pi}{26}\right] + d \, \text{Sec}\left[\frac{5\pi}{26}\right] \right)$

One option now is to consider one coefficient nonzero and all the others zero. For example

FullSimplify$\left[\sum_{k=0}^{\infty}(-1)^k \left(\frac{1}{13k+1} + \frac{1}{13k+12} \right)\right]$
$\frac{2(-1)^{15/26}\pi}{13(-1+(-1)^{2/13})}$

The last number is equal to

$\frac{1}{13}\pi \, \text{Csc}\left[\frac{\pi}{13}\right]$

8.1.7.12   Case 12

There exists a nice series in this case, namely

FullSimplify$\left[\sum_{k=0}^{\infty}(-1)^k \left(\frac{3}{14k+1} + \frac{-3}{14k+3} + \frac{3}{14k+5} + \frac{4}{14k+7} + \frac{3}{14k+9} + \frac{-3}{14k+11} + \frac{3}{14k+13} \right)\right]$
π

We consider now the series

$$\sum_{k=0}^{\infty}(-1)^k \left(\frac{a}{14k+1} + \frac{b}{14k+2} + \frac{c}{14k+3} + \frac{d}{14k+4} + \frac{e}{14k+5} + \frac{f}{14k+6} + \frac{g}{14k+7} + \frac{f}{14k+8} + \frac{e}{14k+9} + \frac{d}{14k+10} + \frac{c}{14k+11} + \frac{b}{14k+12} + \frac{a}{14k+13} \right);$$

FullSimplify[%]

$$\frac{1}{28}\pi \left(g + 2 \left(a\, Csc\left[\frac{\pi}{14}\right] + b\, Csc\left[\frac{\pi}{7}\right] + c\, Csc\left[\frac{3\pi}{14}\right] + f\, Sec\left[\frac{\pi}{14}\right] + e\, Sec\left[\frac{\pi}{7}\right] + d\, Sec\left[\frac{3\pi}{14}\right]\right)\right)$$

We simplify the last sum

FullSimplify[%/.{b→0,d→0,f→0}]

$$\frac{1}{28}\pi \left(g + 2 \left(a\, Csc\left[\frac{\pi}{14}\right] + c\, Csc\left[\frac{3\pi}{14}\right] + e\, Sec\left[\frac{\pi}{7}\right]\right)\right)$$

Because of

FullSimplify$\left[\frac{1}{28} \left(4 + 2 \left(3Csc\left[\frac{\pi}{14}\right] - 3Csc\left[\frac{3\pi}{14}\right] + 3Sec\left[\frac{\pi}{7}\right]\right)\right)\right]$

1

we get the first series.

Remark In [71, (35)] the following series is exhibited

$$\sum_{k=0}^{\infty}(-1)^k \left(\frac{3}{14k+1} + \frac{-3}{14k+3} + \frac{3}{14k+5} + \frac{4}{14k+7} + \frac{4}{14k+9} + \frac{4}{14k+11} + \frac{4}{14k+13} \right);$$

and it is stated that it converges toward π. This is false since

N[%]
3.74057

8.1.7.13 Case 13

Clear[a,b,c,d,e,f,g,h]

$$\sum_{k=0}^{\infty}(-1)^k \left(\frac{a}{16k+1} + \frac{b}{16k+2} + \frac{c}{16k+3} + \frac{d}{16k+4} + \frac{e}{16k+5} + \frac{f}{16k+6} + \frac{g}{16k+7} + \frac{h}{16k+8} + \frac{g}{16k+9} + \frac{f}{16k+10} + \frac{e}{16k+11} + \frac{d}{16k+12} + \frac{c}{16k+13} + \frac{b}{16k+14} + \frac{a}{16k+15} \right);$$

FullSimplify[%]

$$\frac{1}{32}\pi \left(h + 2 \left(\sqrt{2}d + a\, Csc\left[\frac{\pi}{16}\right] + b\, Csc\left[\frac{\pi}{8}\right] + c\, Csc\left[\frac{3\pi}{16}\right] + g\, Sec\left[\frac{\pi}{16}\right] + e\, Sec\left[\frac{3\pi}{16}\right] + 2\sqrt{2}f\, Sin\left[\frac{\pi}{8}\right]\right)\right)$$

The simplest case appears whenever $h = 1$ and $a = b = c = d = e = f = g = 0$. Then we have

$$\frac{32 \sum_{k=0}^{\infty} \frac{(-1)^k}{16k+8}}{\pi}$$

Thus, we get the Gregory and Leibniz formula.

Another particular case consists in $d = 1$ and $a = b = c = e = f = g = h = 0$. In this case we get a series already discussed at Case 2.

Suppose that $g = 1$ and all other coefficients are null. Then we have the following series

$\mathsf{FullSimplify}\left[\sum_{k=0}^{\infty}(-1)^k\left(\frac{1}{16k+7} + \frac{1}{16k+9}\right)\right]$

$\frac{1}{8}\pi\,\mathsf{Csc}\left[\frac{\pi}{8}\right]\mathsf{Sin}\left[\frac{\pi}{16}\right]$

The last result coincides with

$\frac{1}{8}\dfrac{\sqrt{2+\sqrt{2}}\sqrt{2-\sqrt{2+\sqrt{2}}}}{\sqrt{2}}\pi$

8.1.7.14 Case 14

We have the result

$\sum_{k=0}^{\infty}(-1)^k\left(\frac{a}{18k+1} + \frac{b}{18k+3} + \frac{c}{18k+5} + \frac{d}{18k+7} + \frac{d}{18k+11} + \frac{c}{18k+13} + \frac{b}{18k+15} + \frac{a}{18k+17}\right);$

$\mathsf{FullSimplify}[\%]$

$\frac{1}{18}\pi\left(2b + a\,\mathsf{Csc}\left[\frac{\pi}{18}\right] + d\,\mathsf{Sec}\left[\frac{\pi}{9}\right] + c\,\mathsf{Sec}\left[\frac{2\pi}{9}\right]\right)$

Since

$\mathsf{FullSimplify}\left[\mathsf{Csc}\left[\frac{\pi}{18}\right] - \mathsf{Sec}\left[\frac{\pi}{9}\right] + \mathsf{Sec}\left[\frac{2\pi}{9}\right]\right]$
6

we conclude that

$\sum_{k=0}^{\infty}(-1)^k\left(\frac{2}{18k+1} + \frac{3}{18k+3} + \frac{2}{18k+5} + \frac{-2}{18k+7} + \frac{-2}{18k+11} + \frac{2}{18k+13} + \frac{3}{18k+15} + \frac{2}{18k+17}\right);$

$\mathsf{FullSimplify}[\%]$

π

The convergence of the previous series is suggested by the following.

$\mathsf{case14}[\mathsf{n\_}]:=\sum_{k=0}^{n}(-1)^k\left(\frac{2}{18k+1} + \frac{3}{18k+3} + \frac{2}{18k+5} + \frac{-2}{18k+7} + \frac{-2}{18k+11} + \frac{2}{18k+13} + \frac{3}{18k+15} + \frac{2}{18k+17}\right);$

$\mathsf{Table}[\mathsf{N}[\mathsf{case14}[\mathsf{n}]],\{\mathsf{n},6\}$ (* The first six terms of the sequence *)
$\mathsf{DiscretePlot}[\mathsf{case14}[\mathsf{n}],\{\mathsf{n},50\},$
$\mathsf{Ticks}\rightarrow\{\mathsf{Automatic},\{3.12,\pi,3.163\}\},$
$\mathsf{PlotStyle}\rightarrow\mathsf{Black},\mathsf{ImageSize}\rightarrow250]$
$\mathsf{FullSimplify}[\mathsf{case14}[\infty]]$
$\{3.00528, 3.23335, 3.07251, 3.19696, 3.09541, 3.1812\}$

See Fig. 8.9.

8.1.7.15 Case 15

We have the known result

Fig. 8.9 Case 14

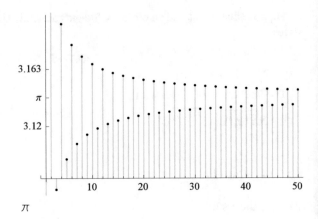

$$\sum_{k=0}^{\infty}(-1)^k \left(\frac{3}{22k+1} + \frac{-3}{22k+3} + \frac{3}{22k+5} + \frac{-3}{22k+7} + \frac{3}{22k+9} + \frac{8}{22k+11} + \frac{3}{22k+13} + \frac{-3}{22k+15} + \frac{3}{22k+19} + \frac{-3}{22k+19} + \frac{3}{22k+21}\right);$$
FullSimplify[%]
π

Remark In [71, (37)] is exhibited the following series

$$\sum_{k=0}^{\infty}(-1)^k \left(\frac{3}{22k+1} + \frac{-3}{22k+3} + \frac{3}{22k+5} + \frac{-3}{22k+7} + \frac{3}{22k+9} + \frac{8}{22k+11} + \frac{3}{22k+13} + \frac{-3}{22k+15} + \frac{3}{22k+19} + \frac{-3}{22k+19} + \frac{1}{22k+21}\right);$$

and it is stated that it converges toward π. This is false since

N[%]
3.075

8.1.7.16 A Case

$$3+4\sum_{k=1}^{\infty} \frac{(-1)^{k+1}}{2k(2k+1)(2k+2)}$$
π

8.1.7.17 A Case

$$6\sum_{k=1}^{\infty} \frac{1}{k^2}$$
π^2

This is exactly the ζ function of Riemann evaluated at 2, i.e., $\zeta(2)$.

8.1.7.18 A Case

$$\frac{8\sum_{k=1}^{\infty} \frac{1}{(2k-1)^2}}{\pi^2}$$

8.2 Newton's Geometric Construction

In [71, (18)] the next definite integral is presented

3 Sqrt[3]/4+24$\int_0^{1/4} \sqrt{x(1-x)}$ dx
Simplify[%]
$$\frac{\frac{3\sqrt{3}}{4} + \frac{1}{4}\left(-3\sqrt{3} + 4\pi\right)}{\pi}$$

8.3 Euler Series

From [71, (23)] we see that

$$2\left\{\sum_{k=0}^{\infty} \frac{2^k k(k!)^2}{(2k+1)!}, \sum_{k=0}^{\infty} \frac{k!}{(2k+1)!!}\right\}$$
$$\{\pi, \pi\}$$

8.4 π by Arcsin

From [71, (26)] we see that

Series$\left[\frac{\text{ArcSin}[x]}{\sqrt{1-x^2}}, \{x, 0, 10\}\right]$
$$\sum_{k=0}^{\infty} \frac{(2x)^{2k+1} k!^2}{2(2k+1)!}$$
%/.x→1/2
$$x + \frac{2x^3}{3} + \frac{8x^5}{15} + \frac{16x^7}{35} + \frac{128x^9}{315} + O[x]^{11}$$
$$\frac{\frac{\text{ArcSin}[x]}{\sqrt{1-x^2}}}{\frac{\pi}{3\sqrt{3}}}$$

8.5 π by the Golden Ratio

f[x_]:=$\frac{\text{ArcSin}[x]}{\sqrt{1-x^2}}$;
f[1/2]

FullSimplify[f[Sin[Pi/10]]]

$\% == \dfrac{\text{Pi}}{5\sqrt{\text{GoldenRatio}+2}}$ (* This is $[71, (28)]$ *)

$\dfrac{\pi}{3\sqrt{3}}$

$\dfrac{1}{5}\sqrt{\dfrac{2}{5+\sqrt{5}}}\,\pi$

True

$\text{GoldenRatio} == \dfrac{1+\sqrt{5}}{2}$

FullSimplify[TrigToExp[FunctionExpand[

$\dfrac{1}{2}\text{Sum}\left[\dfrac{(k!)^2}{\text{GoldenRatio}^{2k+1}(2k+1)!},\{k,0,\text{Infinity}\}\right]]]]$

$N\left[\dfrac{\text{Pi}}{5\sqrt{\text{GoldenRatio}+2}}\right] == N[\%]$

True

$\dfrac{1}{5}\pi\,\text{Root}\left[1-5\#1^2+5\#1^4\&,3\right]$

True

$\text{FullSimplify}\left[\dfrac{1}{5}\sqrt{\dfrac{1}{10}\left(5-\sqrt{5}\right)}-\dfrac{1}{5\sqrt{\text{GoldenRatio}+2}}\right]$

0

$\phi = \text{GoldenRatio};$

$\pi == \text{FullSimplify}\left[\dfrac{5\sqrt{\phi+2}}{2}\sum_{k=0}^{\infty}\dfrac{(k!)^2}{\phi^{2k+1}(2k+1)!}\right]$

True

$\text{FullSimplify}\left[\sum_{k=0}^{\infty}\dfrac{(k!)^2}{\phi^{2k+1}(2k+1)!}\right]$

$== \text{FullSimplify}\left[\dfrac{1}{\phi}\sum_{k=0}^{\infty}\dfrac{1}{\phi^{2k}(2k+1)\text{Binomial}[2k,k]}\right]$

$== \text{FullSimplify}\left[\dfrac{1}{\phi}\sum_{k=0}^{\infty}\dfrac{2k+1}{\phi^{2k}(2k+1)}\int_0^1 x^k(1-x)^k\ dx\right]$

$== \text{FullSimplify}\left[\dfrac{1}{\phi}\int_0^1\left(\sum_{k=0}^{\infty}\left(\dfrac{x}{\phi}\times\dfrac{1-x}{\phi}\right)^k\right)\ dx\right]$

$== \text{FullSimplify}\left[\phi\int_0^1\dfrac{1}{\phi^{2-x(1-x)}}\ dx\right]$

$\dfrac{1}{5}\sqrt{2-\dfrac{2}{\sqrt{5}}}\,\pi == \dfrac{1}{5}\sqrt{2-\dfrac{2}{\sqrt{5}}}\,\pi == \dfrac{1}{5}\sqrt{2-\dfrac{2}{\sqrt{5}}}\,\pi$

$== \dfrac{1}{5}\sqrt{2-\dfrac{2}{\sqrt{5}}}\,\pi == \dfrac{2\text{GoldenRatio}\,\pi}{5\sqrt{5+2\sqrt{5}}}$

The last equality is true because of

$\text{FullSimplify}\left[\dfrac{2\pi}{5\sqrt{5+2\sqrt{5}}}\dfrac{1+\sqrt{5}}{2}\right]$

$\dfrac{1}{5}\sqrt{2-\dfrac{2}{\sqrt{5}}}\,\pi$

We also have

Fig. 8.10 π by the golden ratio

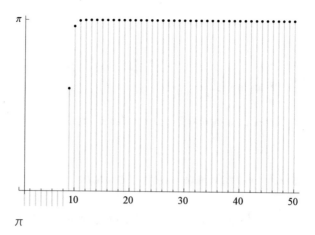

$$\underset{\pi}{\text{FullSimplify}}\left[\frac{5\phi\sqrt{\phi+2}}{2}\int_0^1 \frac{1}{\phi^2-x(1-x)}\,dx\right]$$

The convergence of the first series is suggested by the following code.

```
φ=GoldenRatio;
pibygoldenratio[n_]:=5√φ+2/2 ∑ⁿk=0 (k!)²/φ^(2k+1)(2k+1)!
Table[N[pibygoldenratio[k],10],{k,6}]   (* The first six terms *)
DiscretePlot[pibygoldenratio[n],{n,50},
Ticks→{Automatic,{π}},
PlotStyle→Black,ImageSize→250]
FullSimplify[pibygoldenratio[∞]]
{3.126021252, 3.140314037, 3.141483900, 3.141583199, 3.141591819,
3.141592579}
```

See Fig. 8.10.

8.6 π by Integrals

We recall some integrals connected to the π number.

8.6.1 Dantzell Formula

$$\underset{\pi}{\frac{22}{7}} - \int_0^1 \frac{x^4(1-x)^4}{1+x^2}\,dx \quad (* \ [71, \ (50)] \ *)$$

Hence it immediately follows that $\pi < 22/7$, [41].

8.6.2 Lucas Formula

$$\frac{355}{113} - \frac{1}{3164} \int_0^1 \frac{x^8(1-x)^8(25+816x^2)}{1+x^2} \, dx; \quad (* \ [71, \ (51)] \ *)$$
Simplify[%]
π

Hence it immediately follows that $\pi < 355/113$, [41].

8.6.3 Backhouse-Lucas Formula

In [5, 42], and [71] the following integral is discussed.

i[m_Integers,n_Integers]:=$\int_0^1 \frac{x^m(1-x)^n}{1+x^2}$ dx
i[m,n]
FullSimplify[Assuming[m>0,i[m,2m]]]
i[m,n]
i[m,2 m]

On the same line we mention four more results from [42].

$$\left\{ \int_0^1 \left(\frac{x^{10}(1-x)^8}{4(1+x^2)} + \frac{5}{138450312} \right) \ dx == \frac{355}{113} - \pi, \right.$$
$$\int_0^1 \left(\frac{8192}{114291} \times \frac{x^9(1-x)^8}{\sqrt{1-x^2}} + \frac{15409}{219772564011} \right) \ dx == \frac{355}{113} - \pi,$$
$$\int_0^1 \frac{x^{14}(1-x)^{12}(124360+77159x^2)}{755216(1+x^2)} \ dx == \pi - \frac{103993}{33102},$$
$$\left. \int_0^1 \frac{x^{12}(1-x)^{12}(1349-1060x^2)}{38544(1+x^2)} \ dx == \frac{104348}{33215} - \pi \right\}$$
{True, True, True, True}

8.7 BBP and Adamchik-Wagon Formulas

This section is mainly based on [2, 3, 6], and [69]. A *BBP formula* allows to compute the n-th decimal of π without knowing its previous decimals. The acronym BBP comes from the names of three mathematicians called (David) Bailey, (Peter) Borwein, and (Simon) Plouffe.

An interesting result of this kind is proven in [2].

We underline that the next result, called an Adamchik-Wagon formula (AW formula), is true for any complex number r.

FullSimplify[PowerExpand [
$$\sum_{k=0}^\infty \frac{1}{16^k} \left(\frac{4+8r}{8k+1} - \frac{8r}{8k+2} - \frac{4r}{8k+3} - \frac{2+8r}{8k+4} - \frac{1+2r}{8k+5} - \frac{1+2r}{8k+6} + \frac{r}{8k+7} \right)]]$$
π

We can approach the previous identity by means of the next two commands

```
f[i_]:=2^{i/2}Integrate[Sum[z^(8k + i − 1), {k, 0, Infinity}], {z, 0, 1/Sqrt[2]}]
FullSimplify[PowerExpand[
(4+8r)f[1]-8r f[2]-4r f[3]-(2+8r)f[4]-(1+2r)f[5]-(1+2r)f[6]+r f[7]]]
π
```

The last command can be written under the form

```
FullSimplify[PowerExpand[
{4+8r,-8r,-4r,-(2+8r),-(1+2r),-(1+2r),r}.Array[f,7]]]
π
```

Setting $r = 0$ yields the original BBP formula, [6]

```
FullSimplify[PowerExpand[4f[1]-2f[4]-f[5]-f[6]]]
π
```

Setting $r = -1/2$ yields the BBP formula [71, (39)]

```
FullSimplify[PowerExpand[{0,4,2,2,0,0,-1/2}.Array[f,7]]]
π
```

8.8 A Method for Finding AW Formulas and Proofs

In this section we refer to [2] and [3].

```
as=Array[a,7];
asArray=FullSimplify[TrigToExp[as.Array[f,7]]];
LogExpand[exp_]:=PowerExpand[exp]/.Log[n_Integer]:→
Apply[#2.Log[#1]&,Transpose[FactorInteger[n]]]
GetTranscendentals[exp_]:=Union[Cases[exp,Pi | _ArcTan | _Log | _
ArcCoth,Infinity]]
simpler=Together[LogExpand[asArray]/.{
ArcCot[Sqrt[2]]→Pi/2-ArcTan[Sqrt[2]],ArcCoth[Sqrt[2]]→
Log[Sqrt[2]+1],ArcCoth[4]→Log[5]-Log[3]}];
transsimpler=GetTranscendentals[simpler]
```

$$\left\{\pi, \text{ArcTan}[2], \text{ArcTan}\left[\sqrt{2}\right], \text{Log}[3], \text{Log}[5], \text{Log}\left[1 + \sqrt{2}\right]\right\}$$

```
collected=Map[Factor,Collect[simpler,transsimpler]];
system=DeleteCases[Flatten[CoefficientList[collected,
Append[transsimpler,Sqrt[2]]]],0];
```

```
TableForm[system,7];
(coeffMatrix=Normal[CoefficientArrays[system,as]][[2]])//MatrixForm
MatrixRank[coeffMatrix]
```

$$\begin{pmatrix} \frac{1}{4} & 0 & \frac{1}{2} & 0 & 1 & 0 & 2 \\ \frac{1}{8} & 0 & -\frac{1}{4} & 1 & -\frac{1}{2} & 0 & 1 \\ 0 & \frac{1}{4} & 0 & -1 & 0 & 1 & 0 \\ -\frac{1}{4} & 0 & \frac{1}{2} & 0 & -1 & 0 & 2 \\ \frac{1}{4} & -\frac{1}{2} & \frac{1}{2} & 0 & -1 & 2 & -2 \\ 0 & \frac{1}{4} & 0 & 0 & 0 & -1 & 0 \\ \frac{1}{8} & 0 & -\frac{1}{4} & 0 & \frac{1}{2} & 0 & -1 \end{pmatrix}$$

6

```
sol=Solve[system=={0,0,0,0,0,1,0}]
```

$$\left\{ \left\{ a[2] \to 4 - a[1], a[3] \to 2 - \frac{a[1]}{2}, a[4] \to 1 - \frac{a[1]}{2}, a[5] \to -\frac{a[1]}{4}, \right. \right.$$
$$\left. \left. a[6] \to -\frac{a[1]}{4}, a[7] \to -\frac{1}{2} + \frac{a[1]}{8} \right\} \right\}$$

Let $a[1]$ be a free parameter denoted $8r + 4$. Then

```
a[1]=8r+4;
solution=Simplify[as/.First[sol]]
```
$$\{4 + 8r, -8r, -4r, -1 - 4r, -1 - 2r, -1 - 2r, r\}$$

This yields the AW formula for π at the beginning of the Sect. 8.7.

If we set r to 0, pop out the BBP coefficients

```
solution/.r→0
```
$$\{4, 0, 0, -1, -1, -1, 0\}$$

and thus the BBP formula is proven.

8.9   BBP-Type Formulas for π in Powers of 2^k

We list the formulas (38)–(43) in [71].

$$\text{FullSimplify}\Big[\Big\{\sum_{k=0}^{\infty} \frac{1}{16^k}\Big(\frac{4}{8k+1} - \frac{2}{8k+4} - \frac{1}{8k+5} - \frac{1}{8k+6}\Big),$$

$$\frac{1}{2}\sum_{k=0}^{\infty} \frac{1}{16^k}\Big(\frac{8}{8k+2} + \frac{4}{8k+3} + \frac{4}{8k+4} - \frac{1}{8k+7}\Big),$$

$$\text{FunctionExpand}\Big[\frac{1}{16}\sum_{k=0}^{\infty} \frac{1}{256^k}\Big(\frac{64}{16k+1} - \frac{32}{16k+4} - \frac{16}{16k+5} - \frac{16}{16k+6} + \frac{4}{16k+9} -$$
$$\frac{2}{16k+12} - \frac{1}{16k+13} - \frac{1}{16k+14}\Big)\Big],$$

$$\text{FunctionExpand}\Big[\frac{1}{32}\sum_{k=0}^{\infty} \frac{1}{256^k}\Big(\frac{128}{16k+2} + \frac{64}{16k+3} + \frac{64}{16k+4} - \frac{16}{16k+7} + \frac{8}{16k+10} +$$
$$\frac{4}{16k+11} + \frac{4}{16k+12} - \frac{1}{16k+15}\Big)\Big],$$

$$\frac{1}{32}\sum_{k=0}^{\infty} \frac{1}{4096^k}\Big(\frac{256}{24k+2} + \frac{192}{24k+3} - \frac{256}{24k+4} - \frac{96}{24k+6} - \frac{96}{24k+8} + \frac{16}{24k+10} -$$
$$\frac{4}{24k+12} - \frac{3}{24k+15} - \frac{6}{24k+16} - \frac{2}{24k+18} - \frac{1}{24k+20}\Big),$$

Fig. 8.11 BBP-type formulas for π in powers of 2^k

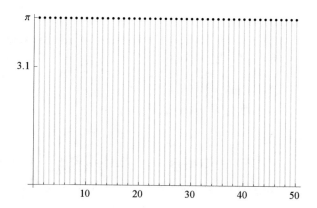

$$\text{TrigToExp}\left[\text{FunctionExpand}\left[\frac{1}{64}\sum_{k=0}^{\infty}\frac{1}{4096^k}\left(\frac{256}{24k+1}+\frac{256}{24k+2}-\frac{384}{24k+3}-\right.\right.\right.$$
$$\frac{256}{24k+4}-\frac{64}{24k+5}+\frac{96}{24k+8}+\frac{64}{24k+9}+\frac{16}{24k+10}+\frac{8}{24k+12}-\frac{4}{24k+13}+\frac{6}{24k+15}+$$
$$\left.\left.\left.\frac{6}{24k+16}+\frac{1}{24k+17}+\frac{1}{24k+18}-\frac{1}{24k+20}-\frac{1}{24k+21}\right)\right]\right]$$

$\{\pi,\pi,\pi,\pi,\pi,\pi\}$

The last series converges rapidly. This fact is suggested by the following results.

$$\text{bbptype}[n\_]:=\frac{1}{64}\sum_{k=0}^{n}\frac{1}{4096^k}\left(\frac{256}{24k+1}+\frac{256}{24k+2}-\frac{384}{24k+3}-\frac{256}{24k+4}-\frac{64}{24k+5}+\right.$$
$$\frac{96}{24k+8}+\frac{64}{24k+9}+\frac{16}{24k+10}+\frac{8}{24k+12}-\frac{4}{24k+13}+\frac{6}{24k+15}+\frac{6}{24k+16}+\frac{1}{24k+17}+$$
$$\left.\frac{1}{24k+18}-\frac{1}{24k+20}-\frac{1}{24k+21}\right)$$

Table[N[bbptype[n],25],{n,6}]   (* The first six terms of the sequence *)
DiscretePlot[bbptype[n],{n,1,50},Ticks→{Automatic,{3.1,π}},
PlotStyle→Black,ImageSize→250,AxesOrigin→{0,3}]
{3.14159265353810642757166650, 3.141592653589789141720046,
3.14159265358979323802020210, 3.141592653589793238462586,
3.14159265358979323846264, 3.1415926535897932384626643}

See Fig. 8.11.

8.9.1 A Case

Another series of the same kind is formula (48) in [71] and we introduce it bellow.

$$\text{sum24}=\frac{1}{96}\sum_{k=0}^{\infty}\frac{1}{4096^k}\left(\frac{256}{24k+2}+\frac{64}{24k+3}+\frac{128}{24k+5}+\frac{352}{24k+6}+\frac{64}{24k+7}+\frac{288}{24k+8}+\right.$$
$$\frac{128}{24k+9}+\frac{80}{24k+10}+\frac{20}{24k+12}-\frac{16}{24k+14}-\frac{1}{24k+15}+\frac{6}{24k+16}-\frac{2}{24k+17}-\frac{1}{24k+19}+$$
$$\left.\frac{1}{24k+20}-\frac{2}{24k+21}\right);$$

FullSimplify[FunctionExpand[%]];

Because of

```
FullSimplify[20ArcCot[4-Sqrt[3]]+20ArcCot[4+Sqrt[3]]]
FullSimplify[-16 Sqrt[3]ArcCot[5/Sqrt[3]]-4 Sqrt[3]ArcCot[7/Sqrt[3]]
-4 Sqrt[3]ArcCot[3 Sqrt[3]]+12 Sqrt[3]ArcCot[11 Sqrt[3]]];
```

$$20\mathrm{ArcTan}\left[\tfrac{2}{3}\right]$$

$$- 2\sqrt{3}\left(\pi + \mathrm{ArcTan}\left[\tfrac{15\sqrt{3}}{337}\right]\right)$$

one has

$$\tfrac{1}{18}\left(21\pi - 4\mathrm{Sqrt}[3]\pi - \mathrm{ArcTan}\left[\tfrac{13651680}{815616479}\right] - 4\mathrm{Sqrt}[3]\mathrm{ArcTan}\left[\tfrac{15\sqrt{3}}{337}\right] + \right.$$
$$16(-1)^{1/12}\sqrt{2}\left((-1)^{5/6}\mathrm{ArcTan}\left[\tfrac{1}{11}\left(4 - 3\sqrt{3}\right)\right] + \right.$$
$$\left.\left.\mathrm{ArcTan}\left[\tfrac{1}{11}\left(-4 + 3\sqrt{3}\right)\right] + \left(-i + (-1)^{1/3}\right)\mathrm{ArcTan}\left[\tfrac{1}{11}\left(4 + 3\sqrt{3}\right)\right]\right)\right);$$

```
sum24a=ComplexExpand[%]
```

$$\tfrac{7\pi}{6} - \tfrac{2\pi}{3\sqrt{3}} - \tfrac{1}{18}\mathrm{ArcTan}\left[\tfrac{13651680}{815616479}\right] - \frac{2\mathrm{ArcTan}\left[\tfrac{15\sqrt{3}}{337}\right]}{3\sqrt{3}} + $$
$$\tfrac{4}{9}\mathrm{ArcTan}\left[\tfrac{1}{11}\left(-4 - 3\sqrt{3}\right)\right] - \frac{4\mathrm{ArcTan}\left[\tfrac{1}{11}\left(-4-3\sqrt{3}\right)\right]}{3\sqrt{3}} - $$
$$\tfrac{4}{9}\mathrm{ArcTan}\left[\tfrac{1}{11}\left(4 - 3\sqrt{3}\right)\right] - \frac{4\mathrm{ArcTan}\left[\tfrac{1}{11}\left(4-3\sqrt{3}\right)\right]}{3\sqrt{3}} + $$
$$\tfrac{4}{9}\mathrm{ArcTan}\left[\tfrac{1}{11}\left(-4 + 3\sqrt{3}\right)\right] + \frac{4\mathrm{ArcTan}\left[\tfrac{1}{11}\left(-4+3\sqrt{3}\right)\right]}{3\sqrt{3}} - $$
$$\tfrac{4}{9}\mathrm{ArcTan}\left[\tfrac{1}{11}\left(4 + 3\sqrt{3}\right)\right] + \frac{4\mathrm{ArcTan}\left[\tfrac{1}{11}\left(4+3\sqrt{3}\right)\right]}{3\sqrt{3}} + $$
$$i\left(-\tfrac{2}{9}\mathrm{Log}\left[1 + \tfrac{1}{121}\left(-4 - 3\sqrt{3}\right)^{2}\right] + \frac{2\mathrm{Log}\left[1+\tfrac{1}{121}\left(-4-3\sqrt{3}\right)^{2}\right]}{3\sqrt{3}} + \right.$$
$$\tfrac{2}{9}\mathrm{Log}\left[1 + \tfrac{1}{121}\left(4 - 3\sqrt{3}\right)^{2}\right] + \frac{2\mathrm{Log}\left[1+\tfrac{1}{121}\left(4-3\sqrt{3}\right)^{2}\right]}{3\sqrt{3}} - $$
$$\tfrac{2}{9}\mathrm{Log}\left[1 + \tfrac{1}{121}\left(-4 + 3\sqrt{3}\right)^{2}\right] - \frac{2\mathrm{Log}\left[1+\tfrac{1}{121}\left(-4+3\sqrt{3}\right)^{2}\right]}{3\sqrt{3}} + $$
$$\left.\tfrac{2}{9}\mathrm{Log}\left[1 + \tfrac{1}{121}\left(4 + 3\sqrt{3}\right)^{2}\right] - \frac{2\mathrm{Log}\left[1+\tfrac{1}{121}\left(4+3\sqrt{3}\right)^{2}\right]}{3\sqrt{3}}\right)$$

We now show that the above number is real and equals π. Firstly we show that its imaginary part is zero.

```
FullSimplify[Im[sum24a]]
0
```

We pick up the real part of sum24a.

FullSimplify[TrigExpand[Re[sum24a]]]

$\frac{1}{18} \left(\left(21 - 4\sqrt{3}\right) \pi - \text{ArcTan}\left[\frac{13651680}{815616479}\right] - 4\sqrt{3}\text{ArcTan}\left[\frac{15\sqrt{3}}{337}\right] + \right.$

$16 \left(\left(1 + \sqrt{3}\right) \text{ArcTan}\left[\frac{1}{11}\left(-4 + 3\sqrt{3}\right)\right] + \right.$

$\left. \left(\left(-1 + \sqrt{3}\right) \text{ArcTan}\left[\frac{1}{11}\left(4 + 3\sqrt{3}\right)\right]\right)\right)$

And now the last step of calculations shows that

N[%-Pi]

0.

We conclude that sum24 $= \pi$.

8.10 π Formulas by Binomial Sums

We introduce three formulas by binomial sums, called in some references *BBP like binomial sums*. For this kind of sums, a great help is offered by the identity

$$\frac{1}{\binom{m \cdot n}{p \cdot n}} = (m \cdot n + 1) \int_0^1 x^{p \cdot n}(1 - x)^{(m-p)n} \, \mathrm{d}x, \tag{8.1}$$

[53, p. 277].

Below we introduce three series of this sort. The first series belongs to W. Gosper, whereas the second to S. Plouffe. The third one belongs to Gourevitch and Guillera, [28]. The first two series follow.

$$\pi = \sum_{k=0}^{\infty} \frac{50k - 6}{\text{Binomial}[3k, k]2^k}$$

$$3 + \pi = \sum_{k=1}^{\infty} \frac{k2^k(k!)^2}{(2k)!} = \sum_{k=1}^{\infty} \frac{k2^k}{\text{Binomial}[2k, k]}$$

We transform the right-hand sides substituting the binomial coefficients with the definite integrals given above by (8.1).

FullSimplify $\left[\text{FunctionExpand} \left[\left\{ \int_0^1 \left(\sum_{k=0}^{\infty} \frac{(50k-6)(3k+1)}{2^k} x^k (1 - x)^{2k} \right) \mathrm{d}x, \right. \right. \right.$

$\left. \left. \left. \int_0^1 \left(\sum_{k=1}^{\infty} k2^k(2k + 1)x^k(1-x)^k \right) \mathrm{d}x \right\} \right] \right]$

$\{\pi, 3 + \pi\}$

The Gourevitch-Guillera series follows.

$$\frac{1}{16807} \sum_{k=0}^{\infty} \frac{1}{2^k \mathrm{Binomial}[7k,2k]} \left(\frac{59296}{7k+1} - \frac{10326}{7k+2} - \frac{3200}{7k+3} - \frac{1352}{7k+4} - \frac{792}{7k+5} + \frac{552}{7k+6} \right);$$

Chop@N[%-π]

0

8.11   S. Ramanujan Series

We are referring to an identity of the form given below, [9, 10], and [7].

$$\frac{\sqrt{8}}{9801} \sum_{k=0}^{\infty} \frac{(4k)!}{(k!)^4} \times \frac{1103 + 26390k}{396^{4k}}$$
$$\frac{1}{\pi}$$

Regarding the convergence of this series we note the following.

ramanujan1[n_]:=$\frac{\sqrt{8}}{9801} \sum_{k=0}^{n} \frac{(4k)!}{(k!)^4} \times \frac{1103+26390k}{396^{4k}}$
{Table[N[1/ramanujan1[n],30],{n,4}],   (* The first four terms *)
Table[N[π -1/ramanujan1[n],30],{n,4}]}   (* It clearly appears that at
each step twelve exact decimals are added *)
DiscretePlot[1/ramanujan1[n],{n,50},Ticks→{Automatic,{3.1,π}},
PlotStyle→Black,ImageSize→250,AxesOrigin→{0,3}]
ramanujan1[∞]
{{3.14159265358979387799890582631,
3.14159265358979323846264490657,
3.14159265358979323846264338328,
3.14159265358979323846264338328},
{$-6.3953626244302651021001947563\overline{3}0.*\wedge$-16,
$-5.6824232560139595080810872289\overline{6}30.*\wedge$-24,
$-5.2388962804811045274090241759\overline{3}30.*\wedge$-32,
$-4.9441875792480300122099869073\overline{7}30.*\wedge$-40}}

See Fig. 8.12.
    Another example on the same line is the following, [30, (1)],

$$\frac{1}{\pi} = \frac{1}{4} \sum_{k=0}^{\infty} \frac{6k+1}{4^k} \frac{\Gamma(1/2+k)(\Gamma(1/2))^3}{k!^3}.$$

FullSimplify$\left[\mathrm{FunctionExpand}\left[\frac{1}{4} \sum_{k=0}^{\infty} \frac{6k+1}{4^k} \times \frac{(\mathrm{Gamma}[1/2+k]/\mathrm{Gamma}[1/2])^3}{k!^3} \right] \right]$

$\frac{\pi + \mathrm{EllipticK}\left[\frac{1}{4}(2-\sqrt{3})\right]^2}{\pi^2} - \frac{\mathrm{Gamma}\left[\frac{1}{6}\right]^2 \mathrm{Gamma}\left[\frac{1}{3}\right]^2}{16\sqrt{3}\pi^3}$

The last result is written as

$$\frac{1}{\pi}\left(1 + \frac{\text{EllipticK}\left[\frac{1}{4}(2-\sqrt{3})\right]^2}{\pi} - \frac{\text{Gamma}\left[\frac{1}{6}\right]^2 \text{Gamma}\left[\frac{1}{3}\right]^2}{16\sqrt{3}\pi^2}\right)$$

We substitute $x = 1/6$ in (6.42) of [53] and have

$$\text{Gamma}\left[\frac{1}{6}\right] = \frac{\sqrt{3}}{2^{1/3}\sqrt{\pi}}\text{Gamma}\left[\frac{1}{3}\right]\wedge 2$$

We show that the difference of the terms in the complete elliptic integral and Γ's
is

$$\frac{1}{2^{2/3}} \times \frac{3}{\text{Pi}}\text{Gamma}\left[\frac{1}{3}\right]^6 - 16\sqrt{3}\pi\,\text{EllipticK}\left[\frac{1}{4}\left(2-\sqrt{3}\right)\right]^2;$$

N[%]
0.

The third example is presented below, [30, (2)],

$$\text{FullSimplify}\left[\text{FunctionExpand}\left[\frac{1}{16}\sum_{k=0}^{\infty}\frac{42k+5}{2^{6k}}\frac{(\text{Gamma}[1/2+k]/\text{Gamma}[1/2])^3}{k!^3}\right]\right]$$

$$\text{Chop@N}\left[\% - \frac{1}{\pi}\right]$$

$$\frac{1}{\pi} + \frac{5\text{EllipticK}\left[\frac{1}{16}(8-3\sqrt{7})\right]^2}{4\pi^2} - \frac{5\text{Gamma}\left[\frac{1}{7}\right]^2\text{Gamma}\left[\frac{2}{7}\right]^2\text{Gamma}\left[\frac{4}{7}\right]^2}{64\sqrt{7}\pi^4}$$

0

The fourth series of Ramanujan type is the next one

Fig. 8.12 An S. Ramanujan
series

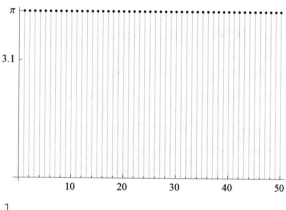

$$\frac{1}{\pi} = \frac{1}{32} \sum_{k=0}^{\infty} \left(\left(42\sqrt{5} + 30 \right) k + 5\sqrt{5} - 1 \right) \times$$

$$\frac{(\text{Gamma}[1/2+k]/\text{Gamma}[1/2])^3}{k!^3} \times \frac{1}{2^{6k}} \left(\frac{\sqrt{5}-1}{2} \right)^{8k}$$

FullSimplify[FunctionExpand[

$$\frac{1}{32} \sum_{k=0}^{\infty} \left(\left(42\sqrt{5} + 30 \right) k + 5\sqrt{5} - 1 \right) \times$$

$$\frac{(\text{Gamma}[1/2+k]/\text{Gamma}[1/2])^3}{k!^3} \times \frac{1}{2^{6k}} \left(\frac{\sqrt{5}-1}{2} \right){}^\wedge(8k)]]$$

Chop@N $\left[\% - \frac{1}{\pi} \right]$

$$\frac{1}{(267-119\sqrt{5})\pi^2} 2\text{EllipticK} \left[\frac{1}{32} \left(16 - 7\sqrt{3} - \sqrt{15} \right) \right]$$

$$\left(2\sqrt{6 \left(27 + 7\sqrt{5} \right) \left(-125 + 56\sqrt{5} \right)} \text{ EllipticE} \left[\frac{1}{32} \left(16 - 7\sqrt{3} - \sqrt{15} \right) \right] +$$

$$\left(328 - 267\sqrt{15} + \sqrt{5 \left(234319 - 35224\sqrt{15} \right)} \right)$$

EllipticK $\left[\frac{1}{32} \left(16 - 7\sqrt{3} - \sqrt{15} \right) \right]$)

0

A short series of Ramanujan follows, [59],

$$\sum_{k=0}^{\infty} (-1)^k (4k + 1) \times \frac{(\text{Gamma}[1/2+k]/\text{Gamma}[1/2])^3}{k!^3};$$

Chop@N $\left[\% - \frac{2}{\pi} \right]$

0

8.12   R. W. Gosper Series

Gosper proved the following identity, [71, (57)],

$$\pi + 4\text{ArcTan}[z] + 2\text{Log} \left[\frac{1-2z-z^2}{z^2+1} \right] =$$

$$\sum_{k=0}^{\infty} \frac{1}{16^k} \left(\frac{4(z+1)^{8k+1}}{8k+1} - \frac{2(z+1)^{8k+4}}{8k+4} - \frac{(z+1)^{8k+5}}{8k+5} - \frac{(z+1)^{8k+6}}{8k+6} \right)$$

Let us see how can one prove it.

$$\sum_{k=0}^{\infty} \frac{1}{16^k} \left(\frac{4(z+1)^{8k+1}}{8k+1} - \frac{2(z+1)^{8k+4}}{8k+4} - \frac{(z+1)^{8k+5}}{8k+5} - \frac{(z+1)^{8k+6}}{8k+6} \right);$$

FullSimplify[FunctionExpand[%]]

$$2 \left(\text{ArcCot} \left[\frac{2}{(1+z)^2} \right] + (1 + i)\text{ArcCot} \left[\frac{1+i}{1+z} \right] +$$

$$(1 + i)\text{ArcCoth} \left[\frac{1+i}{1+z} \right] - \text{Log} \left[\frac{1+\frac{1}{2}(1+z)^2}{\sqrt{1-\frac{1}{4}(1+z)^4}} \right] - \text{Log} \left[\frac{1+\frac{1}{4}(1+z)^4}{\sqrt{1-\frac{1}{16}(1+z)^8}} \right] \right)$$

Define

Fig. 8.13 A domain of
convergence for a Gosper
series

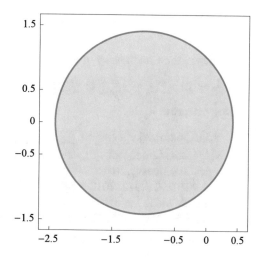

$f1[z\_]:=Pi + 4ArcTan[z] + 2Log\left[\frac{1-2z-z^2}{z^2+1}\right];$

$f2[z\_]:=2\left(ArcCot\left[\frac{2}{(1+z)^2}\right] + (1 + i)ArcCot\left[\frac{1+i}{1+z}\right] +\right.$

$\left.(1 + i)ArcCoth\left[\frac{1+i}{1+z}\right] - Log\left[\frac{1+\frac{1}{2}(1+z)^2}{\sqrt{1-\frac{1}{4}(1+z)^4}}\right] - Log\left[\frac{1+\frac{1}{4}(1+z)^4}{\sqrt{1-\frac{1}{16}(1+z)^8}}\right]\right);$

We show that the derivatives of the two functions coincide and the functions
coincide at a value of the argument.

{FullSimplify[f1'[z]],FullSimplify[f2'[z]]}
{f1[0],FullSimplify[ComplexExpand[f2[0]]]}
$\left\{\frac{16z}{-1+2z+2z^3+z^4}, \frac{16z}{-1+2z+2z^3+z^4}\right\}$
$\{\pi, \pi\}$

We find the domain of convergence of the series.

Clear[k]
SumConvergence
$\left[\frac{1}{16^k}\left(\frac{4(z+1)^{8k+1}}{8k+1} - \frac{2(z+1)^{8k+4}}{8k+4} - \frac{(z+1)^{8k+5}}{8k+5} - \frac{(z+1)^{8k+6}}{8k+6}\right), k\right]$
$Abs[1 + z]^8 < 16$

Let us note that the point $z = 0$ belongs to the domain of convergence.

RegionPlot[Abs[1+x+I y]^2 <2,{x,-2.6,.6},{y,-1.6,1.6},
FrameTicks→{{{-1.5,-.5,0,.5,1.5},None},{{-2.5,-1.5,-.5,0,.5},None}},
ImageSize→200]

See Fig. 8.13.

8.13   The Chudnovskys' Series

The series of the Chudnovsky brothers is as follows [17]:

$$\frac{1}{\pi} = 12 \sum_{k=0}^{\infty} \frac{(-1)^k (6k)!}{(k!)^{\wedge}3(3k)!} \times \frac{13591409+545140134k}{(640320)^{\wedge}(3k+3/2)}$$

Now we prove it.

seriesChudnovsky[n_]:= $12 \sum_{k=0}^{n} \frac{(-1)^k (6k)!}{(k!)^{\wedge}3(3k)!} \times \frac{13591409+545140134k}{(640320)^{\wedge}(3k+3/2)}$
N[1/{seriesChudnovsky[0],seriesChudnovsky[1],
seriesChudnovsky[2]}-Pi,30]
{−5.9030794189791701204543434936130.*^-14,
    3.0784780427869694662126849468030.*^-28,
− 1.7205202595330622168019371734330.*^-42}

Its remarkable property consists in the fact that, with each additional term, about 14 exact digits are added to the sum.

FullSimplify@FunctionExpand@seriesChudnovsky@∞
$\frac{1}{\pi}$

A series given at [9] has a similar remarkable property that each additional term in the series adds about 24 exact digits.

Clear[a,b,chudnov,c]
a=212175710912$\sqrt{61}$ + 1657145277365;
b=13773980892672$\sqrt{61}$ + 107578229802750;
c=5280$\left(236674 + 30303\sqrt{61}\right)$;

chudnov[n_]:=$12\sum_{k=0}^{n} \frac{(-1)^k (6k)!(a+k\ b)}{k!^3 (3k)! c^{3k+3/2}}$;
Block[{MaxExtraPrecision=1000},N[chudnov[#]-$\frac{1}{\pi}$,30]]&/@{0,1,2,3}
Chop@N[FullSimplify[FunctionExpand[chudnov[∞]]]-$\frac{1}{\pi}$]
{1.6128378181616637866979172192730.*^-25,
− 1.4210316904270604870355440806430.*^-50,
    1.3376839203027799643784247246330.*^-75,
− 1.3088943494223346582625533206230.*^-100}
0

8.14   B. Cloitre Series

An interesting sum was found by Cloitre, sum in which the golden ratio is present. The result appeared in [18] and follows below.

g:=GoldenRatio;
$50 \sum_{k=0}^{\infty} \frac{1}{g^{5k}} \left(\frac{g^{-2}}{(5k+1)^2} - \frac{g^{-1}}{(5k+2)^2} - \frac{g^{-2}}{(5k+3)^2} + \frac{g^{-5}}{(5k+4)^2} + \frac{2g^{-5}}{(5k+5)^2} \right)$
π^2

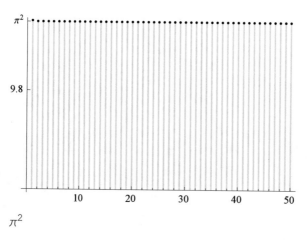

Fig. 8.14 A Cloitre series

Some details regarding the convergence of the series are given below.

cloitre[n_]:=50 $\sum_{k=0}^{n}$ $\frac{1}{g^{5k}}$ $\left(\frac{g^{-2}}{(5k+1)^2} - \frac{g^{-1}}{(5k+2)^2} - \frac{g^{-2}}{(5k+3)^2} + \frac{g^{-5}}{(5k+4)^2} + \frac{2g^{-5}}{(5k+5)^2} \right)$
Table[N[cloitre[n],10],{n,5}]   (* The first five terms *)
Table[N[π^2-cloitre[n]],{n,10}]
{9.870515488, 9.869648052, 9.869606822, 9.869604548, 9.869604411}
{−0.000911086, −0.0000436505, −2.4206409001692464*^-6,
− 1.473621864533925*^-7, −9.56164747378807*^-9,
− 6.49704290367481*^-10, −4.5712766905126045*^-11,
− 3.304023721284466*^-12, −2.4158453015843406*^-13,
− 1.7763568394002505*^-14}

It appears that each additional term in the series adds about one exact digit.

DiscretePlot[cloitre[n],{n,50},Ticks→{Automatic,{9.8,π^2}},
PlotStyle→Black,ImageSize→250,AxesOrigin→{0,9.7}]
cloitre[∞]   (* The sum of the series *)

See Fig. 8.14.

8.15 F. Bellard Series

Fabrice Bellard in [8] has introduced the following rapidly convergent series.

FullSimplify[FunctionExpand[
$\frac{1}{64}$ $\sum_{k=0}^{\infty}$ $\frac{(-1)^k}{2^{10k}}$ $\left(\frac{-32}{4k+1} - \frac{1}{4k+3} + \frac{256}{10k+1} - \frac{64}{10k+3} - \frac{4}{10k+5} - \frac{4}{10k+7} + \frac{1}{10k+9} \right)$]]
π

Let us see the behaviour of the first partial sums of this series.

Fig. 8.15 A Bellard series

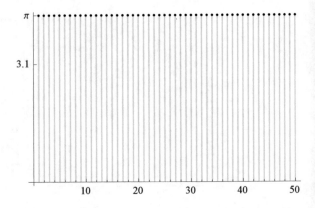

bellard[n_]:=

$$\frac{1}{64}\sum_{k=0}^{n}\frac{(-1)^k}{2^{10k}}\left(\frac{-32}{4k+1}-\frac{1}{4k+3}+\frac{256}{10k+1}-\frac{64}{10k+3}-\frac{4}{10k+5}-\frac{4}{10k+7}+\frac{1}{10k+9}\right)$$

Table[N[bellard[n],15],{n,5}]   (* The first five terms of the sequence *)
Table[N[π-bellard[n],10],{n,10}]   (* At each step about three exact
decimals are added *)
DiscretePlot[bellard[n],{n,50},
Ticks→{Automatic,{3.1,π}},
PlotStyle→Black,
ImageSize→250,
AxesOrigin→{0,3}]
{3.14159257186839, 3.14159265364205, 3.14159265358976,
3.14159265358979, 3.14159265358979}
{8.1721402932088590879196039048411`10.*^-8,
− 5.2257531481640402394406709950451`10.*^-11,
3.7870382130388501392061352355401`10.*^-14,
− 2.9380732247867117491251494295671`10.*^-17,
2.3792815101278340047806579798211`10.*^-20,
− 1.9843261247437912864071983290271`10.*^-23,
1.6908194113321112577208297080601`10.*^-26,
− 1.4644409049327835547169976309631`10.*^-29,
1.2847617481109365037788677210271`10.*^-32,
− 1.1388670796190216333705764169791`10.*^-35}

See Fig. 8.15.

Chapter 9
Optimization of Trajectories

9.1 Necessary Conditions for a Mayer Optimal Control Problem

For brevity, below we introduce an abbreviated version of Theorem 4.2.i in [15, 48]. Let the *Mayer problem* of optimal control be expressed as

$$\Lambda[x, u] = g(a, x(a), b, x(b)) \quad \text{(cost functional)}, \tag{9.1}$$

$$\begin{cases} \frac{\mathrm{d}x}{\mathrm{d}t} = f(t, x(t), u(t)), \quad t \in [a, b], \quad \text{a.e., (differential constraint)} \\ e[x] = (a, x(a), b, x(b)) \in B \subset \mathbb{R}^{1+n+1+n} \quad \text{(boundary conditions)}, \\ (t, x(t)) \in A, \quad t \in [a, b] \quad \text{(time and state constraints)}, \\ u(t) \in U(t), \quad t \in [a, b] \quad \text{(control constraint)}. \end{cases} \tag{9.2}$$

A pair $(x(t), u(t))$, $a \le t \le b$, is said to be *admissible* (or *feasible*) provided that $x : [a, b] \to \mathbb{R}^n$ is absolutely continuous, [47, 53], $u : [a, b] \to \mathbb{R}^p$ is measurable, and x and u satisfy (9.2) a.e. Let Ω be the class of admissible pairs (x, u). The goal is to find the minimum of the cost functional (9.1) over Ω, that is, to find an element $(x^*, u^*) \in \Omega$ so that $-\infty < \Lambda[x^*, u^*] \le \Lambda[x, u]$ for all $(x, u) \in \Omega$. We introduce the variables $\lambda = (\lambda_1, \ldots, \lambda_n)$, called *multipliers*, and an auxiliary function $H(t, x, u, \lambda)$, called the *Hamiltonian*, defined on $T \times U \times \mathbb{R}^n$ by

$$H(t, x, u, \lambda) = \sum_{i=1}^{n} \lambda_i f_i(t, x, u). \tag{9.3}$$

Let us define

$$M(t, x, \lambda) = \inf_{u \in U(t)} H(t, x, u, \lambda).$$

© Springer International Publishing AG 2017
M. Mureşan, *Introduction to Mathematica® with Applications*,
DOI 10.1007/978-3-319-52003-2_9

Further necessary assumptions:

1. There exists an element $(x^*, u^*) \in \Omega$ such that $-\infty < \Lambda[x^*, u^*] \leq \Lambda[x, u]$ for all $(x, u) \in \Omega$.
2. A is closed in $\mathbb{R}^{1+n}$.
3. The set $S = \{(t, x, u) | (t, x) \in A,\ u \in U(t)\}$ is closed in $\mathbb{R}^{1+n+p}$.
4. $f \in C^1(S, \mathbb{R})$.
5. Notation:

$$f_{it} = \frac{\partial f_i}{\partial t}, \quad f_{ix_j} = \frac{\partial f_i}{\partial x_j}, \quad H_{x_j} = \frac{\partial H}{\partial x_j} = \sum_{i=1}^{n} \lambda_i f_{ix_j},$$

$$H_t = \frac{\partial H}{\partial t} = \sum_{i=1}^{n} \lambda_i f_{it}, \quad H_{\lambda_j} = \frac{\partial H}{\partial \lambda_j} = f_j.$$

6. The graph $\{(t, x^*(t)) | a \leq t \leq b\}$ of the optimal trajectory x^* belongs to the interior of A.
7. U does not depend on time and is a closed set.
8. The endpoint $e(x^*) = (a, x^*(a), b, x^*(b))$ of the optimal trajectory x^* is a point of B, where B has a tangent variety B' (of some dimension δ, $0 \leq \delta \leq 2n + 2$) whose vectors are denoted by

$$h = (\tau_1, \xi_1, \tau_2, \xi_2), \quad \xi_1 = (\xi_1^1, \dots, \xi_1^n), \quad \xi_2 = (\xi_2^1, \dots, \xi_2^n),$$

or by

$$h = (da, dx_1, db, dx_2), \quad dx_1 = (d\xi_1^1, \dots, d\xi_1^n), \quad dx_2 = (d\xi_2^1, \dots, d\xi_2^n).$$

Theorem 9.1 *Assume the above eight hypotheses and let (x^*, u^*) be an optimal pair for the Mayer problem* (9.1) *and* (9.2). *Then the optimal pair* (x^*, u^*) *necessarily has the following properties:*

(a) *There exists an absolutely continuous function* $\lambda(t) = (\lambda_1(t), \dots, \lambda_n(t))$ *such that*

$$\frac{d\lambda_i}{dt} = -H_{x_i}\left(t, x^*(t), u^*(t), \lambda(t)\right), \quad i = 1, \dots, n, \quad t \in [a, b] \ (a.e.).$$

If dg *is not identically zero at* $e[x^*]$, *then* $\lambda(t)$ *is never zero in* $[a, b]$.

(b) *For almost any fixed* $t \in [a, b]$ (a.e.), *the Hamiltonian, as a function depending only on* u, *takes its minimum value in* U *at the optimal strategy* $u^* = u^*(t)$. *This implies* $M(t, x^*(t), \lambda(t)) = H(t, x^*(t), u^*(t), \lambda(t))$, $t \in [a, b]$ (a.e.).

(c) *The function* $M(t) = M(t, x^*(t), \lambda(t))$ *coincides a.e. in* $[a, b]$ *with an absolutely continuous function, and*

$$\frac{dM}{dt} = \frac{d}{dt} M\left(t, x^*(t), \lambda(t)\right) = H_t\left(t, x^*(t), u^*(t), \lambda(t)\right), \quad t \in [a, b] \quad (a.e.).$$

(d) *(transversality relation) There exists a constant $\lambda_0 \geq 0$ such that*

$$\left(\lambda_0 g_a - M(a)\right) da + \sum_{i=1}^{n} \left(\lambda_0 g_{x_1^i} + \lambda_i(a)\right) dx_1^i +$$

$$\left(\lambda_0 g_b + M(b)\right) db + \sum_{i=1}^{n} \left(\lambda_0 g_{x_2^i} - \lambda_i(b)\right) dx_2^i = 0, \tag{9.4}$$

for every vector $h = (da, dx_1, db, dx_2) \in B'$.

(e) *(Legendre–Clebsch) If $H(t, x(t), u(t), \lambda(t))$ has second-order partial derivatives with respect to $u_1, \ldots, u_m$ at least in a neighborhood of $u^*(t)$, then we necessarily have that*

$$\sum_{j,k=1}^{m} H_{u_j u_k} \xi_j \xi_k \geq 0$$

for all $\xi = (\xi_1, \ldots, \xi_m) \in \mathbb{R}^m$, all $a < t < b$, where the derivatives $H_{u_j u_k}$ are computed at $(t, x^*(t), u^*(t), \lambda(t))$.

9.2 Zermelo's Navigation Problem

Zermelo's navigation problem requires determination of the optimal trajectory and the associated guidance of a boat (ship, aircraft) traveling between two given points so that the transit time is minimized. In this section, we study this minimum time optimal control problem mainly, numerically, by the power of *Mathematica*. By the Manipulate command, we show the families of trajectories of the navigation problem for certain values of parameters.

9.2.1 Introduction

According to [21, p. 150], Zermelo was the first to formulate and solve in [78] and [79] a problem that now is called the navigation problem of Zermelo. The problem came to Zermelo's mind when the airship Graf Zeppelin circumnavigated the Earth in August 1929. He considered a vector field given in the Euclidean plane that describes the distribution of winds as depending on place and time and treated the question of how an airship or ship, moving at a constant speed against the surrounding air, has to fly in order to reach a given point B from a given point A in the shortest possible time. With:

- $x = x(t)$ and $y = y(t)$ the Cartesian coordinates of the airship at time t,
- $u = u(t, x, y)$ and $v = v(t, x, y)$ the corresponding components of the vector field representing the velocity of the wind (water) with respect to the Cartesian system,
- $\beta = \beta(t, x, y)$ the angle between the momentary speed (u_0, v_0) of the airship against the surrounding air and the x-axis and normalizing to $\| (u_0, v_0) \| = 1$,

one has the system of differential equations that describes the problem

$$\frac{dx}{dt} = u + \cos \beta \quad \text{and} \quad \frac{dy}{dt} = v + \sin \beta.$$

Using the calculus of variations, Zermelo obtained the following differential equation for the heading angle β:

$$\frac{d\beta}{dt} = \sin^2 \beta \cdot \frac{\partial v}{\partial x} + \sin \beta \cos \beta \left(\frac{\partial u}{\partial x} - \frac{\partial v}{\partial y} \right) - \cos^2 \beta \cdot \frac{\partial u}{\partial y}.$$

The previous differential equation, called the Zermelo's differential equation, is a necessary condition for β to be the optimal guidance function.

An extensive list of references on Zermelo's navigation problem and on methods of its study may be found in [56].

Since the navigation problem is obviously nonlinear, up to the author's knowledge, there is no solution in closed form, and its approaches make intensive use of numerical methods. Hereafter, we use the power of *Mathematica* to study the navigation problem of Zermelo numerically.

9.2.2   A Planar Form of the Navigation Problem

Zermelo's navigation problem is a minimum time paths through a region of position-dependent time vector velocity. Under a general form, the problem supposes that a boat travels through a zone of currents. The magnitude and direction of the currents are given by the functions of time and position.

$$u = u(t, x, y) \quad \text{and} \quad v = v(t, x, y)$$

where (x, y) are Cartesian coordinates giving the position of the boat at time t, and (u, v) are the velocity components of the boat at the current point (x, y) at time t in the x and y directions, respectively. The magnitude of the velocity of the boat relative to the water is supposed to be a constant $V > 0$.

The problem requires to steer the boat in such a way as to minimize the time necessary to travel from a given point $A = (x_1, y_1)$ at instant a to another given point $B = (x_2, y_2)$ at instant b. The equations of motion are

$$\begin{cases} x'(t) = V \cos \beta(t) + u(t, x(t), y(t)) \\ y'(t) = V \sin \beta(t) + v(t, x(t), y(t)), \quad t \in [a, b], \end{cases}$$

where β is the heading angle of the boat's axis relative to a fixed coordinate axis. Let the fixed coordinate axis be the horizontal axis and β the control function.

In a more compact form, the navigation problem can be stated as follows:

$$\begin{cases} x'(t) = V \cos \beta(t) + u(t, x(t), y(t)), \\ y'(t) = V \sin \beta(t) + v(t, x(t), y(t)), \\ x(a) = x_1, \quad y(a) = y_1, \quad \text{(initial conditions),} \\ x(b) = x_2, \quad y(b) = y_2, \quad \text{(final conditions),} \\ [a, b], \quad \text{(finite horizon),} \\ \beta \in C([a, b], \mathbb{R}), \quad \text{(control function),} \\ u, v : [a, b] \times \mathbb{R} \times \mathbb{R} \longrightarrow \mathbb{R}, \quad \text{(components of the velocity of water),} \\ V, \quad \text{(relative speed of the water),} \\ g(b) = b \longrightarrow \min, \quad \text{(cost functional).} \end{cases} \quad (9.5)$$

We suppose that the functions u and v are continuous in the first variable and of class C^1 in the second and third variables.

By the maximum principle of Pontryagin under the form in (a) and (b) of Theorem 9.1, we have that

$$\lambda_1' = -\frac{\partial H}{\partial x} = -\lambda_1 \frac{\partial u}{\partial x} - \lambda_2 \frac{\partial v}{\partial x}, \quad \lambda_2' = -\frac{\partial H}{\partial y} = -\lambda_1 \frac{\partial u}{\partial y} - \lambda_2 \frac{\partial v}{\partial y}, \quad (9.6)$$

$$0 = \frac{\partial H}{\partial \beta} = V(-\lambda_1 \sin \beta + \lambda_2 \cos \beta) \Longrightarrow \lambda_1 \sin \beta = \lambda_2 \cos \beta, \quad (9.7)$$

where the Hamiltonian of the system is

$$H(t, x, y, \lambda_1, \lambda_2) = \lambda_1(V \cos \beta + u(t, x, y)) + \lambda_2(V \sin \beta + v(t, x, y)) + 1. \quad (9.8)$$

If we solve the system of differential Eqs. (9.5) and (9.6), we find x, y, β, λ_1, and λ_2. By (9.7) and the initial and final conditions, we find the solutions that solve the navigation problem. Based on [44] or [45], we have that there exists such a solution.

Example If the functions u and v do not depend explicitly upon t, that is,

$$\begin{cases} x'(t) = V\cos\beta(t) + u(x(t), y(t)), \\ y'(t) = V\sin\beta(t) + v(x(t), y(t)), \quad t \in [a, b], \end{cases} \tag{9.9}$$

then the problem is autonomous, and therefore we take $a = 0$ as the initial instant and $(0, 0)$ as the final point.

From now on, we suppose that the functions u and v do not depend on t, i.e., Eqs. (9.9) hold. Then the Hamiltonian does not explicitly depend on t, and then $H = $ constant is a prime integral. Because we minimize time, this constant has to be 0. Then from (9.7) we have that $H = 0$. We invoke (9.6) and get

$$\lambda_1 = \frac{-\cos\beta}{V + u\cos\beta + v\sin\beta} \quad \text{and} \quad \lambda_2 = \frac{-\sin\beta}{V + u\cos\beta + v\sin\beta}. \tag{9.10}$$

Substituting (9.10) in (9.6) (or asking for consistency between (9.8) and $\mathrm{d}\left(\partial_\beta H\right)/\mathrm{d}t = 0$), what follows is the Zermelo's navigation formula:

$$\frac{\mathrm{d}\beta}{\mathrm{d}t} = \sin^2\beta\frac{\partial v}{\partial x} + \sin\beta\cos\beta\left(\frac{\partial u}{\partial x} - \frac{\partial v}{\partial y}\right) - \cos^2\beta\frac{\partial u}{\partial y}. \tag{9.11}$$

Now the nonlinear Eqs. (9.9) and (9.11) give the general solution for our navigation problem. If we take into account the initial and final conditions, we get the concrete solution if the data are consistent.

We study a special case considering for the current of water the following functions:

$$u(x, y) = -(V/h)y, \quad v(x, y) = 0, \tag{9.12}$$

where h is a nonzero real number. Now we express the data of the problem as functions depending on the angle β. From (9.11) to (9.12), we write

$$\frac{\mathrm{d}\beta}{\mathrm{d}t} = \frac{V}{h}\cos^2\beta \implies \tan\beta = \tan\beta_f + \frac{V}{h}\left(t - t_f\right), \tag{9.13}$$

where t_f is the final time and β_f is the final angle, both still unknown.

From the second equation of the system we have

$$\frac{\mathrm{d}y}{\mathrm{d}t} = V\sin\beta \implies y = y(\beta) = h\left(\sec\beta - \sec\beta_f\right).$$

Now we take into account the first equation of the system, that is,

$$\frac{dx}{dt} = V\cos\beta - \frac{V}{h}y \Longrightarrow x = x(\beta) =$$

$$-\frac{h}{2}\left[\ln\left(\frac{\sec\beta_f + \tan\beta_f}{\sec\beta + \tan\beta}\right) + (\tan\beta_f - \tan\beta)\sec\beta_f - (\sec\beta_f - \sec\beta)\tan\beta_f\right].$$

The angular limits of the navigation problem, the initial angle β_0, and the final one β_f can be obtained requiring the following system of nonlinear equation to be valid:

$$x(\beta_0) = x_1 \text{ and } y(\beta_0) = y_1. \tag{9.14}$$

Now all the elements of the trajectory are determined, and we pass to the numerical approach.

9.2.3 Numerical Approach to the Navigation Problem

Clearly the initial time is $a = 0$. We choose the initial position at $(x(0), y(0)) = (7.32, -3.727)$. The final position is at the origin $(0, 0)$. Then by (9.14), we have that the initial angle is $105°$, whereas the final angle is $240.004°$. The minimum time to steer the boat from the initial point to the origin by formula (31) is 5.46439. The trajectory of this problem is given in blue, whereas the heading direction vectors appear in red.

```
boatZermelo[vbig_,h_,x0_,y0_,m_]:=Module[
{torig,thetaf,thetaff,theta0,degree0,degreef,t1,t2,t,timef,x,y,n,i},
x[t_]:=
-(h/2)(Sec[thetaf]×(Tan[thetaf]-Tan[t])-Tan[t]×(Sec[thetaf]-Sec[t])+
Log[Divide[Tan[thetaf]+Sec[thetaf],Tan[t]+Sec[t]]]);
y[t_]:=h(Sec[t]-Sec[thetaf]);
sol=FindRoot[
{-(h/2)(Sec[thetaff]×(Tan[thetaff]-Tan[torig])-Tan[torig]×(Sec[thetaff]-
Sec[torig])+
Log[Divide[Tan[thetaff]+Sec[thetaff],Tan[torig]+Sec[torig]]])==x0,
h(Sec[torig]-Sec[thetaff])==y0},{{torig,1.85},{thetaff,4.17}}];
theta0=torig/.sol;   (* Initial angle *)
thetaf=thetaff/.sol;   (* Final angle *)
timef=(h/vbig)(Tan[thetaf]-Tan[theta0]);   (* Final time *)
degree0=(theta0×360)/N[2π];   (* Initial angle *)
degreef=(thetaf×360)/N[2π];   (* Final angle *)
thetaa[i_]:=theta0+i×(thetaf-theta0)/m;   (* Arrow at each thetaa[i]
angle *)
```

```
Show[ParametricPlot[{x[t],y[t]},{t,theta0,thetaf},
PlotStyle→{Blue,Thick}],
Graphics[{{Red,Arrowheads[0.02],
Arrow[Table[{{x[thetaa[n]],y[thetaa[n]]},
{x[thetaa[n]]+vbig×Cos[thetaa[n]],
y[thetaa[n]]+vbig×Sin[thetaa[n]]}},{n,0,m}]]},
{Black,Arrowheads[0.02],Arrow[{{x[thetaa[0]],y[thetaa[0]]},
{x[thetaa[0]]-(vbig/h)y[thetaa[0]],y[thetaa[0]]}}]},
{Blue,Arrowheads[0.025],
Arrow[{{x[thetaa[0]],y[thetaa[0]]},{x[thetaa[0]]-(vbig/h)y[thetaa[0]]+
vbig×Cos[thetaa[0]],y[thetaa[0]]+vbig×Sin[thetaa[0]]}}]},
{PointSize[0.015],Blue,Point[{x[thetaa[0]],y[thetaa[0]]}],
Point[{x[thetaa[m]],y[thetaa[m]]}]},
Text[Style[Row[{Style["minimum time of traveling",Italic]," = ",timef}],13],
{x0/2,y0/3}],
Text[Style[A,Italic,12],{x0-0.3,y0}],Text[Style[O,Italic,12],{-0.25,0.25}],
Text[Style[Row[{Style["from A",Italic]," = (", x0,",",y0,") ",
Style[" to O=(0,0)",Italic]}],13],{x0/2,y0/2}],
Text[Style[Row[{Style["initial angle in A",Italic]," = ",
Superscript[degree0," ∘"]}],13],{x0/2,2y0/3}],
Text[Style[Row[{Style["final angle in O",Italic]," = ",
Superscript[degreef," ∘"]}],13],{x0/2,4y0/4.8}]}],
PlotRange→All,ImageSize→500]]
```

The graph below exhibits the dynamics of the navigation problem when the initial data belong to some intervals. The black (horizontal) vector at *A* represents the direction and magnitude of the current vector at the initial point. The blue (oblique) vector at *A* represents the tangent to the trajectory.

```
Manipulate[
Quiet@boatZermelo[vbig,h,x0,y0,m],
{{vbig,2,V},.5,5,0.5,Appearance→"Labeled"},
{{h,2,h},0.5,5,0.5,Appearance→"Labeled"},
{{x0,7,"x0"},5,20,1,Appearance→"Labeled"},
{{y0,-2,"y0"},-7,3,1,Appearance→"Labeled"},
{{m,15,"number of arrows"},10,30,1,Appearance→"Labeled"},
SaveDefinitions→True
]
```

See Fig. 9.1.

Remark A forthcoming interesting work on the Zermelo's navigation problem is [25].

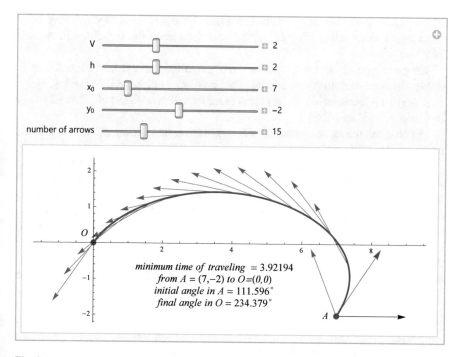

Fig. 9.1 Zermelo's navigation problem

9.3   Optimal Guidance for Planar Lunar Ascent

The present section is focused on the equations of the optimal guidance for planar *Lunar ascent* from rest to insertion in a circular *parking orbit*. The goal is to minimize the flight time under certain assumptions. We point out some computational and graphical aspects for flights reaching two altitudes.

The problem of soft landing a spacecraft on the Moon is a known topic, and we mention [55] and the references therein.

9.3.1   Introduction

The studies on the problem of the ascent from the lunar surface are not new; they started before 1965 as it clearly follows from [14, 46] to [40]. At that time, the methods of calculus of variations were still in force, and more and more strongly the optimal control approach pointed out to be the way to get newer results.

In the lunar ascent problem, the spacecraft is launched from the surface of the moon and ascends to the orbit insertion point, which is defined by a given final velocity vector and a given altitude (sometimes also called free downrange). In spite of the gravity perturbations, it is expected that the ascent trajectory will stay on or close to the plane containing the great circle defined by the launch point and the insertion point.

We point out some computational and graphical aspects of this problem in connection with certain recent results given by D. G. Hull in [35, 36] and [37].

A pertinent introduction on this topic together with a large list of references may be found in [36] and [37].

For constant thrust, minimum time and minimum fuel consumption are the same problem; see [13] or [15].

This section is organized as follows: Sect. 9.3.2 introduces the problem of the ascent from the lunar surface giving some references to make clear the framework of the paper and introduces equations, approximations, and initial values of the ascent problem; Sect. 9.3.3 contains an optimal control approach of determining the optimal trajectory under certain assumptions; and Sect. 9.3.4 discusses some computational and graphical issues of earlier results connected to the problem of determining the optimal trajectory. The Sect. 9.3.4.1 exhibits some figures on the case of 15240 m altitude, whereas the Sect. 9.3.4.2 introduces some figures for the case of 100 km altitude.

9.3.2   The Ascent from the Lunar Surface

Following [36], we consider the ascent problem of a spacecraft from the surface of Moon. The equations of motion in three-dimensional flight over a nonrotating, spherical Moon are written in the local horizontal/local vertical frame as shown in Fig. 9.2. For applications, we considered the data in [36].

The parameters involved in the flight are as follows:

1. $[0, t_f]$ is the horizon of time. The final instant t_f of the insertion on the coast or parking trajectory is to be found in many cases.
2. x and y are the in-plane curvilinear distance (downrange) and altitude. The altitude is specified. More specifically, in Sect. 9.3.4, we discuss in detail the problem for two altitudes of 15240 m and 100 km.
3. u and v are the in-plane horizontal and vertical velocity components. We assume that $v(t_f) = 0$.
4. z and w are the out-of-plane curvilinear distance and velocity components.
5. θ is the thrust pitch angle.
6. ψ is the thrust yaw angle.
7. $r_m = 1738.14 \times 10^3$ m is the radius of the Moon.
8. $g_m = \mu_m / r_m^2 = 1.619 \, \text{m/s}^2$ is the acceleration of gravity on the surface of the Moon, μ_m being the gravitational constant of the Moon.
9. τ is the thrust to mass ratio T/m, where the thrust is $T = 60030 \, N$, and the mass equation is $m(t) = m_0 - \beta t$, $m_0 = 17512.6836 \, \text{kg}$. For constant thrust, $T = \beta V_e$,

Fig. 9.2 Flight over a spherical Moon

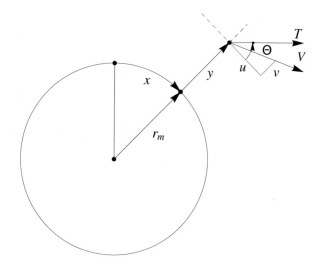

where $\beta = 19.118\,\text{kg/s}$ is the constant propellant mass flow rate, and V_e is the constant exhaust velocity.

From 9 it follows that $\tau = \tau(t) = T/m = T/(m_0 - \beta t) = -V_e/(t - \alpha)$, where $\alpha = m_0/\beta$.

If $z \equiv w \equiv \psi \equiv 0$, then the ascent is said to be *planar*.

The equations of motion for a spacecraft in three-dimensional flight over a non-rotating spherical moon are written in the local horizontal/local vertical coordinate system. Then x, y, u, and v are the in-plane downrange, altitude, horizontal velocity, and vertical velocity component, suggested by Fig. 9.2.

According to [36] and [37], the equations of motion are given by

$$x' = r_m u/r, \quad u' = \tau \cos\theta \cos\psi + (uw/r)\tan(z/r_m) - uv/r,$$

$$y' = v, \quad v' = \tau \sin\theta - g + u^2/r + w^2/r, \tag{9.15}$$

$$z' = r_m w/r, \quad w' = \tau \cos\theta \sin\psi - (u^2/r)\tan(z/r_m) - vw/r.$$

The ascent trajectory from the lunar surface to a coast orbit consists of a (sometimes vertical) rise, constant rate pitchover, and optimal guidance phases. It is designed that the trajectory lies in the plane of the great circle plane containing the launch point and the orbit insertion point. Small deviations from the plane occur because of perturbations on gravity and spacecraft performance. Because the out-of-plane motion is small, the trajectory is said to be *quasiplanar*.

Because the insertion point is relatively close to the lunar surface, we consider the following approximations:

$y/r_m \ll 1$, so that $r \approx r_m$ and $g \approx \mu_m/r_m^2 = g_m$.

The out-of-plane components z and w are assumed to be sufficiently small that the terms z/r_m and w/r_m can also be neglected.

Fig. 9.3 Flight over a flat
Moon

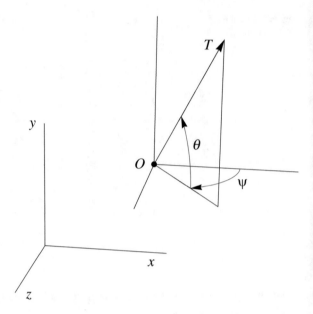

For typical launch trajectories, it can be shown that $uv/r_m \ll T/m$, so the last term in the equation of u' is neglected.

Under these circumstances, the Eq. (9.15) can be written under the simplified form as

$$
\begin{aligned}
x' &= u, & u' &= \tau \cos\theta \cos\psi, \\
y' &= v, & v' &= \tau \sin\theta - g_m, \\
z' &= w, & w' &= \tau \cos\theta \sin\psi.
\end{aligned}
\tag{9.16}
$$

By flight over a flat moon, as in [36], we mean flight over a spherical moon at low altitude and low velocities. While the low altitude condition ($y/r_m \ll 1$) is met so that $r \approx r_m$, $g \approx g_m$, the low velocity requirement ($u^2/(g_m r_m) \ll 1$) is not; see [36]. Flight over a flat moon is suggested by Fig. 9.3.

We note that for flight over a flat moon, the gravity vector is along the radius vector.

9.3.3 The Optimal Control Approach

As we mentioned earlier, for constant thrust, the problem of minimizing fuel consumption is the same as minimizing flight time. Hence, the optimal control problem requires to minimize the performance index (cost functional)

$$
\Lambda = \Lambda[x, y, z, \theta, \psi] = t_f
\tag{9.17}
$$

subject to the differential constraints (9.16), the following initial constraints

$$x(0) = 0, \quad y(0) = 0, \quad z(0) = 0, \quad u(0) = 0, \quad v(0) = 0, \text{ and } w(0) = 0,$$
(9.18)

and the final boundary constraints, as suggested in [35] and [36],

$$y(t_f) = 15240, \quad z(t_f) = 0, \quad u(t_f) = 1625, \quad v(t_f) = 0, \text{ and } w(t_f) = 0.$$
(9.19)

The control variables are the angles $\theta = \theta(t)$ and $\psi = \psi(t)$.
The Hamiltonian of the optimal control problem (9.16)–(9.19) is of the form

$$H = \lambda_1 u + \lambda_2 v + \lambda_3 w + \lambda_4 \tau \cos \theta \cos \psi + \lambda_5 (\tau \sin \theta - g_m) + \lambda_6 \tau \cos \theta \sin \psi.$$

By (a) in Theorem 9.1, we have the first necessary condition under the form of the system of differential equations to the multipliers

$$\lambda_1' = 0, \quad \lambda_2' = 0, \quad \lambda_3' = 0, \quad \lambda_4' = -\lambda_1, \quad \lambda_5' = -\lambda_2, \text{ and } \lambda_6' = -\lambda_3.$$

Immediately, the equations of the multipliers follow:

$$\lambda_1 = c_1, \quad \lambda_2 = c_2, \quad \lambda_3 = c_3,$$
$$\lambda_4 = -c_1 t + c_4, \quad \lambda_5 = -c_2 t + c_5, \quad \lambda_6 = -c_3 t + c_6,$$

where the cs are all constants of integration.
By (b) in Theorem 9.1, the equations of the optimal controls are given by

$$H_\theta = -\lambda_4 \tau \sin \theta \cos \psi + \lambda_5 \tau \cos \theta - \lambda_6 \tau \sin \theta \sin \psi = 0,$$
$$H_\psi = -\lambda_4 \tau \cos \theta \sin \psi + \lambda_6 \tau \cos \theta \cos \psi = 0.$$
(9.20)

Under the necessary condition of ascent, we suppose that $\cos \theta > 0$, and therefore from the latter equation in (9.20), it follows that

$$\lambda_4 \sin \psi - \lambda_6 \cos \psi = 0,$$
(9.21)

and for $|\psi|$ small also $\lambda_4 \tan \psi - \lambda_6 = 0$. Thus, the first equation in (9.20) is written as

$$(\lambda_4 \cos \psi + \lambda_6 \sin \psi) \tan \theta - \lambda_5 = 0.$$
(9.22)

From (9.21) to (9.22), we have the multipliers λ_5 and λ_6 depending on λ_4, that is, $\lambda_6 = \lambda_4 \tan \psi$ and $\lambda_5 = \lambda_4 \tan \theta \sec \psi$.
The second-order partial derivatives of H with respect to the control variables are $H_{\theta\theta} = -\tau \lambda_4 \sec \theta \sec \psi$, $H_{\psi\psi} = -\tau \lambda_4 \cos \theta \sec \psi$, and $H_{\theta\psi} = 0$.

By the Legendre–Clebsch necessary condition in Theorem 9.1 for a minimum, we write $H_{\theta\theta} \geq 0$, $H_{\psi\psi} \geq 0$, and $H_{\theta\theta}H_{\psi\psi} - (H_{\theta\psi})^2 \geq 0$. Then it follows that

$$\text{sgn}(\lambda_4 \cos\theta \cos\psi) \leq 0. \tag{9.23}$$

Thus, necessarily $\lambda_4 \leq 0$.

From (9.21) for some real $\varepsilon = \pm 1$, we write

$$\frac{\varepsilon\lambda_6}{\sqrt{\lambda_4^2 + \lambda_6^2}} \cos\psi - \frac{\varepsilon\lambda_4}{\sqrt{\lambda_4^2 + \lambda_6^2}} \sin\psi = 0.$$

Therefore, we can deduce the equations of the optimal yaw angle

$$\sin\psi = \varepsilon\lambda_6/\sqrt{\lambda_4^2 + \lambda_6^2} \quad \text{and} \quad \cos\psi = \varepsilon\lambda_4/\sqrt{\lambda_4^2 + \lambda_6^2},$$

if $\lambda_4^2 + \lambda_6^2 \neq 0$. Suppose for a while that $\lambda_4 \equiv \lambda_6 \equiv 0$. Then by (9.22), it follows that $\lambda_5 \equiv 0$. Thus, all the Lagrange multipliers are null, and therefore we get a contradiction.

Naturally we admit that $\lambda_4^2 + \lambda_6^2 \neq 0$. From the first equation in (9.20), we successively have for some real $\delta = \pm 1$ that

$$H_\theta = 0 \implies$$

$$-\lambda_4 \sin\theta \frac{\varepsilon\lambda_4}{\sqrt{\lambda_4^2 + \lambda_6^2}} + \lambda_5 \cos\theta - \lambda_6 \sin\theta \frac{\varepsilon\lambda_6}{\sqrt{\lambda_4^2 + \lambda_6^2}} = 0 \implies$$

$$-\varepsilon \frac{\lambda_4^2 + \lambda_6^2}{\sqrt{\lambda_4^2 + \lambda_6^2}} \sin\theta + \lambda_5 \cos\theta = 0 \implies$$

$$-\varepsilon \frac{\delta\sqrt{\lambda_4^2 + \lambda_6^2}}{\sqrt{\lambda_4^2 + \lambda_5^2 + \lambda_6^2}} \sin\theta + \frac{\delta\lambda_5}{\sqrt{\lambda_4^2 + \lambda_5^2 + \lambda_6^2}} \cos\theta = 0.$$

Now we impose condition (9.23) and get that $\delta = -1$. For ε we choose -1 because λ_4 is negative, and for small values of ψ, $\cos\psi$ is positive. From the last equation, we can now deduce the equations of the optimal pitch angle

$$\sin\theta = -\frac{\lambda_5}{\sqrt{\lambda_4^2 + \lambda_5^2 + \lambda_6^2}} \quad \text{and} \quad \cos\theta = \frac{\sqrt{\lambda_4^2 + \lambda_6^2}}{\sqrt{\lambda_4^2 + \lambda_5^2 + \lambda_6^2}}. \tag{9.24}$$

With these results on the expressions of angles, by (9.16), the velocity and acceleration equations of the optimal trajectory, if any, are of the form

$$x' = u, \quad u' = -\tau \frac{\lambda_4}{\sqrt{\lambda_4^2 + \lambda_5^2 + \lambda_6^2}},$$

$$y' = v, \quad v' = -\tau \frac{\lambda_5}{\sqrt{\lambda_4^2 + \lambda_5^2 + \lambda_6^2}} - g_m, \tag{9.25}$$

$$z' = w, \quad w' = -\tau \frac{\lambda_6}{\sqrt{\lambda_4^2 + \lambda_5^2 + \lambda_6^2}}.$$

Lemma 9.1 *The tangential velocity function u is of class C^1, positive and increasing on the interval $[0, t_f]$..*

Proof It is enough to note that u' in (9.25) is continuous and positive. □

Lemma 9.2 *The radial velocity function v is of class C^1 and positive on the interval $[0, t_f]$.*

Following [36] and [37], we introduce the functions A, B, and C by

$$A(t) = A(t, c_1, \ldots, c_6) = \int_0^t \frac{ds}{(s - \alpha)\sqrt{ps^2 + qs + r}},$$

$$B(t) = B(t, c_1, \ldots, c_6) = \int_0^t \frac{s \, ds}{(s - \alpha)\sqrt{ps^2 + qs + r}}, \tag{9.26}$$

$$C(t) = C(t, c_1, \ldots, c_6) = \int_0^t \frac{s^2 \, ds}{(s - \alpha)\sqrt{ps^2 + qs + r}},$$

where the constants p, q, and r are defined as follows:

$$p = c_1^2 + c_2^2 + c_3^2, \quad q = -2(c_1c_4 + c_2c_5 + c_3c_6), \quad \text{and} \quad r = c_4^2 + c_5^2 + c_6^2.$$

Lemma 9.3 *Assume that $t \geq 0$. The following two integral equalities hold*

$$\int_0^t A(s) \, ds = tA(t) - B(t) \quad \text{and} \quad \int_0^t B(s) \, ds = tB(t) - C(t). \tag{9.27}$$

Proof The two integral equalities are obtained by integration by parts; see [53, Chap. 6]. □

Theorem 9.2 *From (9.25) with (9.26) and (9.27) by integration, we get the equations of velocities and states.*

$$u(t) = u(0) - V_e c_1 B + V_e c_4 A,$$
$$v(t) = v(0) - V_e c_2 B + V_e c_5 A - g_m t,$$
$$w(t) = w(0) - V_e c_3 B + V_e c_6 A,$$
$$x(t) = x(0) + t u(t) + V_e c_1 C - V_e c_4 B, \tag{9.28}$$
$$y(t) = y(0) + t v(t) + V_e c_2 C - V_e c_5 B + g_m t^2/2,$$
$$z(t) = z(0) + t w(t) + V_e c_3 C - V_e c_6 B,$$

Remark By Eq. (9.28), we know the form of the optimal trajectories, if any. What has remained is the step of determining the constants of integration cs such that the seeking trajectory satisfies the boundary conditions.

In order to solve the minimum time optimal control problem, let us suppose that all the initial and final values are given, that is, we know $x(0)$, $y(0)$, $z(0)$, $u(0)$, $v(0)$, $w(0)$, $y(t_f)$, $z(t_f)$, $u(t_f)$, $v(t_f)$, and $w(t_f)$. The variables to be found are t_f and $x(t_f)$. Then we write

$$V_x = u(t_f) - u(0), \quad V_y = v(t_f) - v(0) + g_m t_f, \quad V_z = w(t_f) - w(0),$$
$$R_x = x(t_f) - x(0) - t_f u(t_f), \quad R_y = y(t_f) - y(0) - t_f v(t_f) - g_m t_f^2/2, \tag{9.29}$$
$$R_z = z(t_f) - x(0) - t_f w(t_f),$$

and further

$$V_x = - V_e c_1 B(t_f) + V_e c_4 A(t_f), \quad V_y = -V_e c_2 B(t_f) + V_e c_5 A(t_f),$$
$$V_z = - V_e c_3 B(t_f) + V_e c_6 A(t_f),$$
$$R_x = V_e c_1 C(t_f) - V_e c_4 B(t_f), \quad R_y = V_e c_2 C(t_f) - V_e c_5 B(t_f), \tag{9.30}$$
$$R_z = V_e c_3 C(t_f) - V_e c_6 B(t_f).$$

Both sides of the preceding equalities depend on the final instant t_f.

Because t_f is free, by the transversality condition in Theorem 9.1, we have that $H(t_f) = -1$, which gives

$$c_1 u(t_f) + c_2 v(t_f) + c_3 w(t_f) - (-c_2 t_f + c_5) g_m$$
$$- (T/(m_0 - \beta t_f)) \sqrt{(-c_1 t_f + c_4)^2 + (-c_2 t_f + c_5)^2 + (-c_3 t_f + c_6)^2} = -1. \tag{9.31}$$

Equations (9.30) and (9.31) form a system of seven nonlinear equations in the seven unknowns $t_f, c_1, c_2, c_3, c_4, c_5$, and c_6. Since $x(t_f)$ is free, by the transversality condition, we have that $c_1 = 0$, so we only have six unknowns to determine.

Theorem 9.3 *If the nonlinear system formed by the Eqs. (9.30) and (9.31) has a unique system of real solutions, then the minimum time optimum control problem has a unique solution. If the nonlinear system has several systems of real solutions, then we have to check on which one the minimum time is achieved.*

Remark Based on Filippov's existence Theorem for Mayer problem of optimal control as it is given in [15, p. 310], the problem discussed by Theorem (9.3) has a solution.

9.3.4 Computational Issues

This subsection uses the results of the previous one and analyzes their effectiveness.

Remark Denote $X = ax^2 + bx + c$, where $a = p$, $b = 2p\alpha + q$, and $c = p\alpha^2 + q\alpha + r$. We have that

$$P(x) = \int \frac{dx}{x\sqrt{X}} = -\frac{1}{\sqrt{c}} \ln \left| \frac{2c + bx + 2\sqrt{cX}}{x} \right|, \qquad [29, \text{p. } 97]$$

$$Q(t) = \int \frac{dx}{\sqrt{X}} = \frac{1}{\sqrt{a}} \ln \left| \frac{b + 2ax + 2\sqrt{aX}}{\sqrt{b^2 - 4ac}} \right|, \qquad [29, \text{p. } 94]$$

$$R(x) = \int \frac{x\,dx}{\sqrt{X}} = \frac{\sqrt{X}}{a} - \frac{b}{2a} Q(x), \qquad [29, \text{p. } 96].$$

Then the functions A, B, and C are written as

$$A(t) = P(t - \alpha) - P(-\alpha),$$

$$B(t) = Q(t - \alpha) - Q(-\alpha) + \alpha(P(t - \alpha) - P(-\alpha)),$$

$$C(t) = R(t - \alpha) - R(-\alpha) + 2\alpha(Q(t - \alpha) - Q(-\alpha)) + \alpha^2(P(t - \alpha) - P(-\alpha)).$$

These formulas are useful for an easier approach to numerical computation in Theorem 9.3.

Remark In case of the planar flight, we have to consider that $z = w = \psi = 0$. These equalities imply that

$$\lambda_3 = \lambda_6 = 0.$$

Remark Hereafter, we discuss the planar case using the initial and final values as they are given in [35]. In this case, the multipliers are of the form

$$\lambda_1 = c_1, \quad \lambda_2 = c_2, \quad \lambda_3 = -c_1 t + c_3, \quad \text{and} \quad \lambda_4 = -c_2 t + c_4,$$

and the Eqs. (9.28) are written as

$$u(t) = u(0) - V_e c_1 B + V_e c_3 A, \quad v(t) = v(0) - V_e c_2 B + V_e c_4 A - g_m t,$$

$$x(t) = x(0) + tu(t) + V_e c_1 C - V_e c_3 B, \quad (9.32)$$

$$y(t) = y(0) + tv(t) + V_e c_2 C - V_e c_4 B + g_m t^2/2.$$

If we integrate the velocity equations in (9.32), then the equations of states can be written also as

$$x(t) = x(0) + u(0)t + V_e c_3 tA(t) - V_e(c_1 t + c_3)B(t) + V_e c_1 C(t),$$

$$y(t) = y(0) + v(0)t + V_e c_4 tA(t) - V_e(c_2 t + c_4)B(t) + V_e c_2 C(t) - g_m t^2/2.$$

If we denote by (x_c, y_c) the pair of Cartesian components of the (x, y) point in local horizontal/local vertical coordinates, then

$$x_c(t) = (r_m + y(t)) \sin(x(t)/r_m) \text{ and } y_c(t) = (r_m + y(t)) \cos(x(t)/r_m).$$

Immediately it follows that $y_c'(0) = x_c'(0) = 0$.

Lemma 9.4 *We have that*

$$y_c''(0) = \frac{-V_e c_4 - g_m \alpha \sqrt{r}}{\alpha \sqrt{r}} \text{ and } x_c''(0) = \frac{-V_e c_3}{\alpha \sqrt{r}}.$$

Then by the L'Hospital Theorem, [53, p. 207], the angle of the spacecraft at the take-off moment is

$$\arctan \frac{V_e c_4 + g_m \alpha \sqrt{r}}{V_e c_3}. \quad (9.33)$$

9.3.4.1 Ascent to 15,240 m Altitude

In this subsection, we present the numerical results and graphics for the ascent to the 15,240 m altitude.

Remark Consider the data given in Sect. 9.3.2 and the guess (approximate) values as they are listed in [35], i.e.,

$$t_f = 404, \quad c_1 = 0, \quad c_2 = -5.87 \times 10^{-5}, \quad c_3 = -0.0399, \text{ and } c_4 = -0.029.$$

With these initial values, we have obtained the following final moment and coefficients of the Lagrange multipliers:

$$t_f = 393.784, \quad c_1 = 0, \quad c_2 = -2.60297 \times 10^{-4},$$

$$c_3 = -0.170877, \quad \text{and} \quad c_4 = -0.124794.$$

According to our calculation, the downrange in this setting is $x(t_f) = 279,348$.

If we introduce the following constants to measure the degree of accuracy of our calculations

$$\delta_1 = -V_x - V_e(c_1 B(t_f) - c_3 A(t_f)), \quad \delta_2 = -V_y - V_e(c_2 B(t_f) - c_4 A(t_f)),$$

$$\delta_3 = -R_x + V_e(c_1 C(t_f) - c_3 B(t_f)), \quad \delta_4 = -R_y + V_e(c_2 C(t_f) - c_4 B(t_f)),$$

$$\delta_5 = c_1 u(t_f) + c_2 v(t_f) - (-c_2 t_f + c_4) g_m$$

$$- (T/(m_0 - \beta t_f)) \sqrt{(-c_1 t_f + c_3)^2 + (-c_2 t_f + c_4)^2},$$

then the results obtained are given in Table 9.1.

Lemma 9.5 *By (9.24) the angle of the thrust vector at the take-off moment is given by*

$$\lim_{t \downarrow 0} \arctan(-\lambda_4/-\lambda_3) = 36.1412°.$$

Applying (9.33), the angle of the spacecraft in respect to the local horizon at the take-off moment is 8.27585°.

In the last part of the present subsection, we introduce some graphics visualizing the numerical results.

Figures 9.4 and 9.5 introduce the variations of the tangential and the radial velocities.

Table 9.1 Accuracy of computation

$\delta_1 = -6.13909 \times 10^{-12}$	$\delta_2 = 6.13909 \times 10^{-12}$	
$\delta_3 = 2.32831 \times 10^{-9}$	$\delta_4 = -6.69388 \times 10^{-9}$	$\delta_5 = -1.0$

Fig. 9.4 Tangential velocity

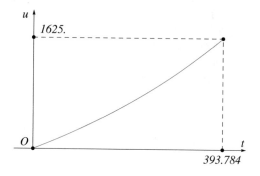

Fig. 9.5 Radial velocity

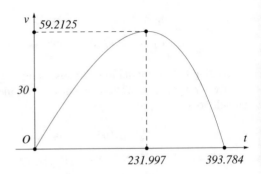

Coast trajectory

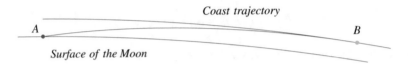

Surface of the Moon

Fig. 9.6 Lunar ascent—complete time

Fig. 9.7 Lunar
ascent—during the first
second

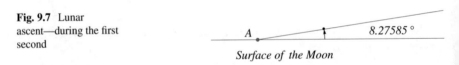

Surface of the Moon

Figure 9.6 represents the whole trajectory of the spacecraft with a part of the
lunar surface. A is the initial point, whereas B is the final point.

The picture that we introduce in Fig. 9.7 represents the trajectory of the spacecraft
constrained on the time interval $t \in [0, 1]$. The parameter of the lunar arc belongs to
the interval $[\pi/2 - 0.0000008, \pi/2 + 0.0000002]$.

Figures 9.4, 9.5, 9.6, and 9.7 are obtained by means of the code which follows:

```
(* Optimal guidance for a planar Lunar ascent *)
Clear[" c*"," t*"," u*"," v*"," x*"," y*" ,m0,beta]
t0=.0;   (* initial time s *)
x0=.0;y0=.0;   (* initial site in local horizontal system of
coordinates *)
u0=.0;v0=.0;   (* initial velocities m/s *)
rm=1738.14×10³;   (* radius of the Moon m *)
gm=1.619;   (* gravitational acceleration on the moon m/s² *)
thrust= 60030.0;   (* thrust N *)
m0=17512.6836;   (* initial mass kg *)
beta=19.118;
```

```
vexit=thrust/beta;
(* xb=279280.0; downrange m *)
yb=15240.0;    (* final altitude m *)
ub=1625.0; vb=0.0;    (* final velocities m/s *)
(* tb=394.0; final time s *)
velx=ub-u0; vely=vb-v0+gm×tb;    (* residues of velocities *)
radx=x[tb]-x0-ub×tb; rady=yb-y0-vb×tb-0.5×gm×tb²;
(* residues of states *)
alpha=m0/beta;
p=a=c1²+c2²;q=-2(c1×c3+c2×c4);r=c4²+c3²;
b=2 p×alpha+q;c=p×alpha²+q×alpha+r;
pbig[x_,c1_,c2_,c3_,c4_]:=-(1/Sqrt[c])×Log[Abs[
Divide[2 c+b×x+2Sqrt[c×(a×x²+b×x+c)],x]]];
qbig[x_,c1_,c2_,c3_,c4_]:=(1/Sqrt[a])×Log[Abs[Divide[
2Sqrt[a×(a× x²+b× x+c)]+2a×x+b,Sqrt[b²-4a×c]]]];
rbig[x_,c1_,c2_,c3_,c4_]:= Divide[Sqrt[a×x²+b×x+c],a]-
Divide[b,2a]×qbig[x,c1,c2,c3,c4];
aa[c1_,c2_,c3_,c4_]:=pbig[tb-alpha,c1,c2,c3,c4]-
pbig[-alpha,c1,c2,c3,c4];
b[c1_,c2_,c3_,c4_]:=qbig[tb-alpha,c1,c2,c3,c4]-
qbig[-alpha,c1,c2,c3,c4]+alpha×(pbig[tb-alpha,c1,c2,c3,c4]-
pbig[-alpha,c1,c2,c3,c4]);
cc[c1_,c2_,c3_,c4_]:=rbig[tb-alpha,c1,c2,c3,c4]-
rbig[-alpha,c1,c2,c3,c4]+2alpha×(qbig[tb-alpha,c1,c2,c3,c4]-
qbig[-alpha,c1,c2,c3,c4])+alpha²×(pbig[tb-alpha,c1,c2,c3,c4]-
pbig[-alpha,c1,c2,c3,c4]);
c1=0.0; (* because of the downrange is free *)
root=FindRoot[{c1×ub+c2×vb+(c2×tb-c4)gm-
(thrust/(m0-beta×tb))×Sqrt[(c1×tb-c3)²+(c2×tb-c4)²]==-1,
velx==-vexit×c1×bb[c1,c2,c3,c4]+vexit×c3×aa[c1,c2,c3,c4],
vely==-vexit×c2×bb[c1,c2,c3,c4]+vexit×c4×aa[c1,c2,c3,c4],
rady==vexit×c2×cc[c1,c2,c3,c4]-vexit×c4×bb[c1,c2,c3,c4]},
{{tb,404.0},{c2,-5.87×10⁻⁵},{c3,-.0399},{c4,-.029}},
AccuracyGoal→10,PrecisionGoal→10];
tb=root[[1,2]];c2=root[[2,2]];c3=root[[3,2]];c4=root[[4,2]];
(* Build up the trajectory *)
aaa[t_]:=pbig[t-alpha,c1,c2,c3,c4]-pbig[-alpha,c1,c2,c3,c4];
bbb[t_]:=qbig[t-alpha,c1,c2,c3,c4]-qbig[-alpha,c1,c2,c3,c4]+
alpha×(pbig[t-alpha,c1,c2,c3,c4]-pbig[-alpha,c1,c2,c3,c4]);
ccc[t_]:=rbig[t-alpha,c1,c2,c3,c4]-rbig[-alpha,c1,c2,c3,c4]+
2alpha×(qbig[t-alpha,c1,c2,c3,c4]-qbig[-alpha,c1,c2,c3,c4])+
alpha²×(pbig[t-alpha,c1,c2,c3,c4]- pbig[-alpha,c1,c2,c3,c4]);
u[t_]:=u0-vexit×(c1×bbb[t]-c3×aaa[t]);
v[t_]:=v0-vexit×(c2×bbb[t]-c4×aaa[t])-gm×t;
```

```
x[t_]:=x0+u[t]×t+vexit×(c1×ccc[t]-c3× bbb[t]);
y[t_]:=y0+v[t]×t+vexit×(c2×ccc[t]-c4× bbb[t])+0.5×gm×t²;
radx==vexit×c1×cc[c1,c2,c3,c4]-vexit× c3× bb[c1,c2,c3,c4];
{value1=-velx-vexit×c1×bb[c1,c2,c3,c4]+vexit×c3×aa[c1,c2,c3,c4],
value2=-vely-vexit×c2×bb[c1,c2,c3,c4]+vexit×c4×aa[c1,c2,c3,c4],
value3=-radx+vexit×c1×cc[c1,c2,c3,c4]-vexit×c3×bb[c1,c2,c3,c4],
value4=-rady+vexit×c2×cc[c1,c2,c3,c4]-vexit×c4×bb[c1,c2,c3,c4],
value5=c1×ub+c2×vb+(c2×tb-c4)gm-
(thrust/(m0-beta×tb))×Sqrt[(c1×tb-c3)²+(c2×tb-c4)²],x[tb]}
xcart[t_]:=(rm+y[t])Sin[Divide[x[t],rm]];
ycart[t_]:=(rm+y[t])Cos[Divide[x[t],rm]];
xinit=xcart[t0];yinit=ycart[t0];
xfinal=xcart[tb];yfinal=ycart[tb];
theta[t_]:=ArcTan[Divide[c2×t-c4,c1×t-c3]]×180/N[Pi];
(* Tangential and Radial velocities *)
{Plot[u[t],{t,t0,tb}],Plot[v[t],{t,t0,tb}],Plot[theta[t],{t,t0,1}]};
roott=t/.Solve[vexit(c2×t-c4)==-gm(t-alpha)Sqrt[p×t²+q×t+r],t][[2]];
vroott=v[roott];
{Show[Plot[u[t],{t,t0,tb}],
Graphics[{{Arrowheads[.02],Arrow[{{-40,0},{tb+50,0}}],
Arrow[{{0,0},{0,u[tb]+400}}]},
{PointSize[0.015],Point[{0,0}],Point[{tb,0}],Point[{0,u[tb]}],
Point[{tb,u[tb]}]},{Dashed,Line[{{tb,0},{tb,u[tb]}}],
Line[{{0,u[tb]},{tb,u[tb]}}]},Text[Style[O,Italic,12],{-18,100}],
Text[Style[t,Italic,12],{tb+43,110}],
Text[Style[u,Italic,12],{-18,u[tb]+340}],
Text[Style[tb,Italic,12],{tb,-140}],
Text[Style[u[tb],Italic,12],{40,u[tb]+120}]}],
PlotRange→All,Axes→False,ImageSize→250],
Show[Plot[v[t],{t,t0,tb}],Graphics[{{Arrowheads[.02],
Arrow[{{-40,0},{tb+50,0}}],Arrow[{{0,0},{0,69}}]},
{PointSize[0.015],Point[{0,0}],Point[{tb,0}],Point[{roott,0}],
Point[{roott,vroott}],Point[{0,vroott}],Point[{0,30}],
Point[{0,vroott}],Point[{tb,v[tb]}]},
{Dashed,Line[{{roott,0},{roott,vroott}}],
Line[{{0,vroott},{roott,vroott}}]},Text[Style[O,Italic,12],{-18,5}],
Text[Style[t,Italic,12],{tb+43,5}],Text[Style[v,Italic,12],{-16,66}],
Text[Style[tb,Italic,12],{tb,-6}],Text[Style[" 30",Italic,12],{-24,30}],
Text[Style[vroott,Italic,12],{50,63}],
Text[Style[roott,Italic,12],{roott,-6}]}],
PlotRange→All,Axes→False,ImageSize→250]}
tau[t_]:=Divide[thrust,m0-beta×t];
vprime[t_]:=-tau[t]×Divide[-c2×t+c4,Sqrt[(c1×t-c3)²+(c2×t-c4)²]]-gm;
uprime[t_]:=-tau[t]×Divide[-c1×t+c3,Sqrt[(c1×t-c3)²+(c2×t-c4)²]];
```

```
{Plot[tau[t],{t,t0,tb}],Plot[vprime[t],{t,t0,tb}]}
ArcTan[(ycart[1]-ycart[0])/(xcart[1]-xcart[0])]×180/N[Pi];
theta[t0];Limit[theta[t],t→0]
Limit[Divide[ycart[t]-ycart[0],t],t→0];
Limit[Divide[xcart[t]-xcart[0],t],t→0];
anglestarting=Divide[vexit×c4+gm×alpha×Sqrt[r],vexit×c3];
anglestartingdegree=ArcTan[Divide[vexit×c4+gm×alpha×Sqrt[r],
vexit×c3]]×180/N[Pi];
anglestartingarctan=ArcTan[anglestarting];
circle[radius_,θ_]:=radius{Cos[θ],Sin[θ]};
{Show[ParametricPlot[{xcart[t],ycart[t]},{t,t0,1}],
ParametricPlot[circle[rm,a],{a,Pi/2-0.0000008,Pi/2+0.0000002}],
Graphics[{PointSize[0.015],{Red,Point[{xinit,yinit}]},
Text[Style[" ",Italic,12],{xinit,yinit+0000}],
Text[Style[A,Italic,12],{xinit-.07,yinit+.05}],
Text[Style["Surface of the Moon",Italic,12],{xinit+.2,yinit-.11}],
Text[Style[" ",Italic,12],{xinit,yinit-0.2}],
Text[Style[anglestartingdegree"°",Italic,12],{xinit+1.,yinit+.06}],
circle[{xinit,yinit},0.5,{0,anglestartingarctan}],
{Arrowheads[.02],Arrow[{{xinit+0.5Cos[anglestartingarctan-0.0001],
yinit+0.5Sin[anglestartingarctan-0.0001]},
{xinit+0.5Cos[anglestartingarctan],yinit+0.5Sin[anglestartingarctan]}
}]}}],
PlotRange→All,Axes→False,ImageSize→250],
radius=Sqrt[xfinal²+yfinal²];
Show[ParametricPlot[{xcart[t],ycart[t]},{t,t0,tb}],
ParametricPlot[circle[rm,a],{a,13.4Pi/30,24.2Pi/48}],
ParametricPlot[circle[radius,a],{a,13.3Pi/30,24.0Pi/48}],
Graphics[{PointSize[0.010],{Red,Point[{xinit,yinit}]},
{Green,Point[{xfinal,yfinal}]},
Text[Style[" ",Italic,12],{xinit,yinit+90000}],
Text[Style[A,Italic,12],{xinit-7000,yinit+7000}],
Text[Style[B,Italic,12],{xfinal,yfinal+11000}],
Text[Style["Coast trajectory",Italic,12],(radius+15000){Cos[Pi/2-0.1],
Sin[Pi/2-0.1]}],
Text[Style["Surface of the Moon",Italic,12],{xinit+50000,yinit-15000}],
Text[Style[" ",Italic,12],{xinit,yinit-70000}],
Text[Style[" ",Italic,12],{xfinal-10000,yfinal+45000}]}],
PlotRange→All,Axes→False,ImageSize→400]}
Show[ParametricPlot[{xcart[t],ycart[t]},{t,t0,tb}],
ParametricPlot[circle[rm,a],{a,0,2Pi}],
Graphics[{PointSize[0.010],{Black,Point[{0,0}]},
{Red,Point[{xinit,yinit}]},{Green,Point[{xfinal,yfinal}]},
Text[Style[" ",Italic,12],{xinit,yinit+130000}],
```

Table 9.2 Accuracy of computation for 100 km altitude

$\delta_1 = -2.27374 \times 10^{-12}$	$\delta_2 = -4.09273 \times 10^{-12}$	
$\delta_3 = 1.04774 \times 10^{-9}$	$\delta_4 = 4.42378 \times 10^{-9}$	$\delta_5 = -1.0$

     Text[Style[A,Italic,18],{xinit,yinit+130000}],
     Text[Style[B,Italic,18],{xfinal,yfinal+85000}],
     Text[Style["Surface of the Moon",Italic,18],{xinit,yinit-210000}],
     Text[Style[" ",Italic,12],{xinit,yinit-70000}],
     Text[Style[" ",Italic,12],{xfinal-10000,yfinal+45000}]}],
     PlotRange→All,Axes→False,ImageSize→400]

9.3.4.2 Ascent to 100,000 m Altitude

In this subsubsection, we present the numerical results and graphics for the ascent to 100,000 m altitude, that is, $y(t_f) = 100,000$. The results and figures of the present case follow the previous code by changing the initial data.

Remark We use the same guess (approximate) values as they are given in the previous subsubsection. We have obtained the following final moment and coefficients of the Lagrange multipliers

$$t_f = 440.185, \quad c_1 = 0, \quad c_2 = -0.000715588,$$

$$c_3 = -0.114418, \quad \text{and} \quad c_4 = -0.244682.$$

According to our calculation, the downrange in this setting is $x(t_f) = 273,612$.

Similar to Table 9.1, we introduce the values of the degree of accuracy in the present setting. The results are given by Table 9.2.

Lemma 9.6 *The angle of the thrust vector at the take-off moment is given by*

$$\theta = 64.9383°,$$

whereas the angle of the spacecraft at the take-off moment is

$$45.6646°.$$

In the sequel similar to the previous subsubsection, we exhibit some figures related to the case of ascent to 100 km altitude.

Figures 9.8 and 9.9 introduce the variations of the tangential and the radial velocities.

Figure 9.10 represents the trajectory of the rocket and the entire lunar surface. The lunar surface is assimilated to a circle.

Fig. 9.8 Tangential velocity;
100 km altitude

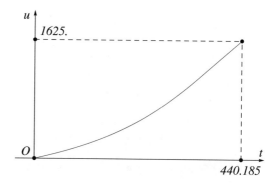

Fig. 9.9 Radial velocity;
100 km altitude

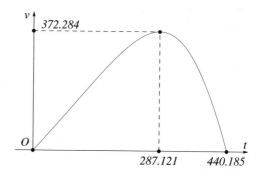

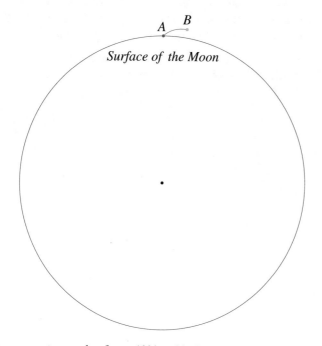

Fig. 9.10 Lunar ascent—complete figure; 100 km altitude

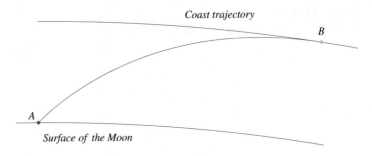

Fig. 9.11 Lunar ascent—complete time; 100 km altitude

Fig. 9.12 Lunar
ascent—during the first
second; 100 km altitude

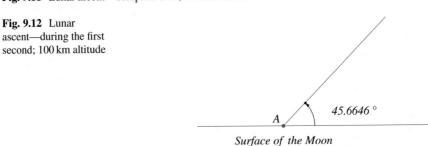

Figure 9.11 represents the whole trajectory of the spacecraft with a part of the lunar surface.

The picture that we introduce in Fig. 9.12 represents the trajectory of the spacecraft constrained on the time interval $t \in [0, 1]$. The parameter of the lunar arc belongs to the interval $[\pi/2 - 0.0000006, \pi/2 + 0.00000035]$.

Remark The maximal orbit transfer problem is connected to the problem of lunar ascent. An approach to the maximal orbit transfer problem with similar method is realized in [57]. Other flying problems are discussed in [48].

Chapter 10
Miscellany in the Euclidean Plane

10.1 Some Planar Curves

In this section we present some applications of *Mathematica* inspired by various sources in planar geometry. Many of the applications are based on the use of the built-in functions Manipulate or Animate.

10.1.1 Conchoid of Nicomedes

The *conchoid of Nicomedes* is the locus of points a fixed distance away from a line as measured along a line from the focus point. The polar equation of the conchoid of Nicomedes is $r(t) = b + a \sec(t)$, whereas in Cartesian coordinates, we write the parametric equations for it $t \longrightarrow (b + a\cos(t))(\cos(t), \sin(t))$. Clearly the function is 2π-periodic. In the graph bellow we set $b = 1$.

```
Clear[a,θ]
GraphicsGrid[With[{eps=10^-5},
Function[a,Show[ParametricPlot[(1+a Sec@θ){Cos@θ,Sin@θ},
Evaluate[{θ,#[[1]]+eps,#[[2]]-eps}],PlotStyle→Red]&/@
Partition[Range[0,2,1/2]π,2,1],PlotRange→{{-2,2},{-2,2}},
Ticks→None,PlotLabel→ToExpression["a"]==a]]/@{0,.03,.3,.5,1,2}]},
ImageSize→600]
```

See Fig. 10.1.

```
ParametricPlot[(1+a Sec@θ){Cos@θ,Sin@θ}/.a→0.003,
{θ,0,2π},ColorFunction→ "Rainbow",ImageSize→100,
Ticks→{{-1,-.5,.5,1},{-3,-2,2,3}}]
```

See Fig. 10.2.

© Springer International Publishing AG 2017
M. Mureşan, *Introduction to Mathematica® with Applications*,
DOI 10.1007/978-3-319-52003-2_10

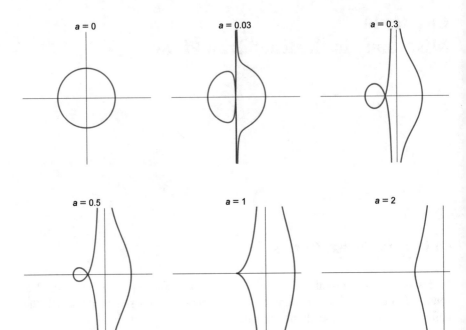

Fig. 10.1 Conchoid of Nicomedes, 1

Fig. 10.2 Conchoid of
Nicomedes, 2

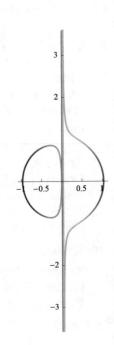

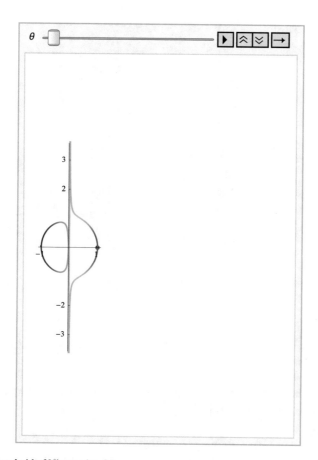

Fig. 10.3 Conchoid of Nicomedes, 3

Below we want to see how a point describes this curve.

```
With[{a=0.03},
point[θ_]:=(1+a Sec[θ]){Cos[θ],Sin[θ]};
conchoid=ParametricPlot[point[θ],{θ,0,2π},
ColorFunction→"Rainbow",ImageSize→100,
Ticks→{{-1,1},{-3,-2,2,3}}]];

Animate[
Show[conchoid,Graphics[{{PointSize[.07],Red,Point[point[θ]]}}],
ImageSize→75],{θ,0,2π},AnimationRate→.025,
AnimationRunning→False,SaveDefinitions→True]
```

See Fig. 10.3.

Fig. 10.4 Conchoid of
Nicomedes, 4

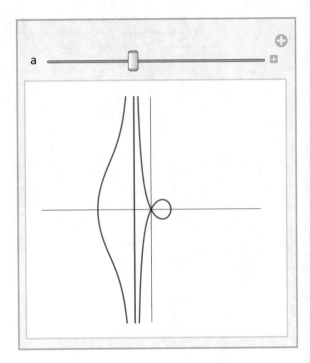

Now we show the evolution of the conchoid under the assumption that the
parameter *a* varies. We set $a \in [-2, 2]$.

```
Clear[a]
conchoidofNicomedes[a_]:=Module[{θ},
Show[ParametricPlot[(1+a Sec[θ]){Cos[θ],Sin[θ]},{θ,0,2 Pi},
PlotStyle→Red,PlotRange→{{-3,3},{-3,3}},ImageSize→200,
Ticks→None]]]

Manipulate[Quiet@conchoidofNicomedes[a],
{{a,-.46,"a"},-2,2,.00001,Appearance→"Labelled"},
SaveDefinitions→True]
```

See Fig. 10.4.

10.1.2 Cycloid

A *cycloid* is the curve described by a point P on the circumference of the circle of
radius $a(> 0)$ which rolls on a straight line. If the straight line is the horizontal real
axis, the parametric equations of P are given by $t \longrightarrow a(t - \sin(t), 1 - \cos(t))$.

Fig. 10.5 Cycloid, 1

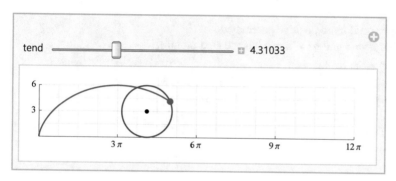

Fig. 10.6 Cycloid, 2

```
Clear[a,x]
x[t_]:=a{t-Sin[t],1-Cos[t]}
ParametricPlot[x[t]/.a→3,{t,0,#},ImageSize→{140,140},
PlotStyle→Red,Ticks→{None}]&/@(π{1/2,1,3/2,2})
```

See Fig. 10.5.

The visualization of the geometric definition of the cycloid follows.

```
cycloid[tend_]:=
Module[{a=3,x,y,t,xx,yy,gridlinex,gridlinesy},
x[t_]:=a(t-Sin[t]); y[t_]:=a(1-Cos[t]);
gridlinesx=Table[{xx,GrayLevel[.5]},{xx,0,4aπ,aπ/4}];
gridlinesy=Table[{yy,GrayLevel[.5]},{yy,0,2a,a/2}];
ParametricPlot[{x[t],y[t]},{t,0,tend},
PlotStyle→{Red,Thick},
Epilog→{Thick,Blue,Circle[a{tend,1},a],
Red,PointSize[Large],Point[{x[tend],y[tend]}],
Black,PointSize[Medium],Point[a{tend,1}]},
GridLines→{gridlinesx,gridlinesy},
Ticks→{aπ Range[4],a{1,2}},PlotRange→2a{{0,2π},{0,1}}]]

Manipulate[Quiet@cycloid[tend],{tend,.0001,4π,
Appearance→"Labeled"},SaveDefinitions→True]
```

See Fig. 10.6.

10.1.3 Epicycloid

An *epicycloid* is the curve described by a point P fixed on the circumference of the circle of radius $b(> 0)$ which rolls on the outside of the circle of radius $a(> 0)$.
Its parametric equations are

$$\begin{cases} x(t) = (a + b)\cos t - b\cos((a/b + 1)t) \\ y(t) = (a + b)\sin t - b\sin((a/b + 1)t), \end{cases}$$

where a and b are positive parameters.
For $a = 8$ and $b = 5$, one has the next picture.

```
Clear[x,y,a,b]
x[t_,p_:1]:=(a+b)Cos[t]-p b Cos[(a/b+1)t]
y[t_,p_:1]:=(a+b)Sin[t]-p b Sin[(a/b+1)t]
ParametricPlot[{x[t],y[t]}/.{a→8,b→5},{t,0,#},
Ticks→{{-15,15},{-15,15}}]&/@(2π{1,2,5,10})
```

See Fig. 10.7.

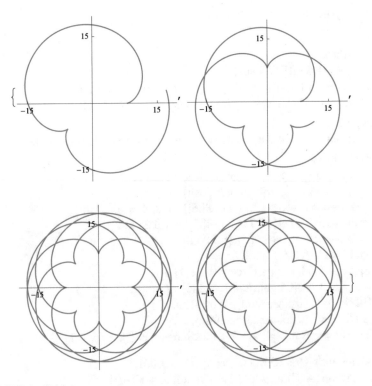

Fig. 10.7 Epicycloid, 1

The visualization of the geometric definition of the epicycloid follows.

```
epicycloid[tend_]:=
Module[{a=3,b=1,x,y,p,xx,gridlines},
x[t_,p_:1]:=(a+b)Cos[t]-p b Cos[(a+b)t/b];
y[t_,p_:1]:=(a+b)Sin[t]-p b Sin[(a+b)t/b];
gridlines=Table[{xx,GrayLevel[0.5]},{xx,-(a+2b),a+2b,(a+2b)/8}];
ParametricPlot[{x[t],y[t]},{t,0,tend},PlotStyle→{Red,Thick},
ImageSize→300,Epilog→{Thick,Blue,Circle[{0,0},a],Black,
Circle[{x[tend,0],y[tend,0]},b],Red,PointSize[Large],
Point[{x[tend],y[tend]}],Black,Line[{{x[tend,0],y[tend,0]},
{x[tend],y[tend]}}]},GridLines→{gridlines,gridlines},
PlotRange→(a+b)(1+1/5){-1,1.1}] ]
```

```
Manipulate[Quiet@epicycloid[tend],{tend,0.001,2π,
Appearance→"Labeled"},SaveDefinitions→True]
```
See Fig. 10.8.

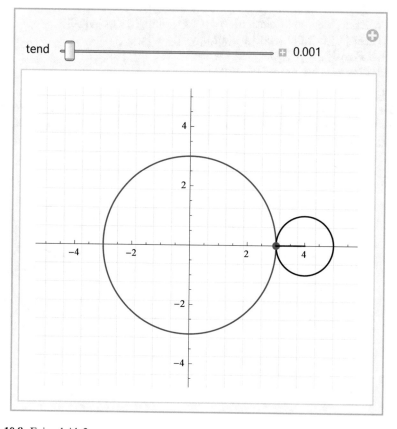

Fig. 10.8 Epicycloid, 2

The proof of the parametric equations of the epicycloid is given here.

```
Clear[a,b,x,y,p]
x[t_,p_:1]:=(a+b)Cos[t]-p b Cos[(a+b)t/b]
y[t_,p_:1]:=(a+b)Sin[t]-p b Sin[(a+b)t/b]
With[{a=3,b=1,t=π/6},
Show[ParametricPlot[{x[tt],y[tt]},{tt,0,t},Axes→None,
PlotStyle→{Red,Thick}],
Graphics[{Circle[{0,0},a],{Blue,Circle[{x[t,0],y[t,0]},b]},
Line[{{{0,0},{x[t,0],y[t,0]}},{a{Cos[t],Sin[t]},a{Cos[t],0}},
{{x[t,0],y[t,0]},{x[t,0],0}},{{x[t,0],y[t,0]},{x[t],y[t]}},
{{x[t],y[t]},{x[t],0}},{{0,0},{a+b,0}},{{0,0},{0,a+b/2}},
{{x[t,0],y[t]},{x[t],y[t]}}}],
{Text[Style[O,Italic,12],{0,-.3}],Text[Style[P,Italic,12],{x[t]+.35,y[t]}],
Text[Style[C,Italic,12],{x[t,0]+.3,y[t,0]}],
Text[Style[D,Italic,12],{a Cos[t],-.3}],Text[Style[T,Italic,12],{x[t],-.3}],
Text[Style[B,Italic,12],{x[t,0],-.3}],Text[Style[S,Italic,12],{a+.2,.2}],
Text[Style[R,Italic,12],{x[t,0]-.22,y[t]+.1}],
Text[Style[Q,Italic,12],{a Cos[t]-.26,a Sin[t]}]},
PointSize[Medium],Point[{{0,0},a{Cos[t],Sin[t]},{x[t,0],y[t,0]},
a{Cos[t],0},{x[t,0],0},{x[t],0},{x[t,0],y[t]},{a,0}}],
Red,Point[{x[t],y[t]}]}],
PlotRange→All,ImageSize→300]]
```

See Fig. 10.9.

Data: $OQ = OS = a$, $CQ = CP = b$, angle$(DOQ) = t$, arc$(SQ) = at = $ arc(QP), $OB = (a+b)\cos t$, $CD = (a+b)\sin t$, $DQ = a\sin t$, $OD = a\cos t$, angle$(QCP) = at/b$, angle$(OCB) = \pi/2 - t$, angle$(BCP) = at/b + t - \pi/2$.

Fig. 10.9 Epicycloid, 3

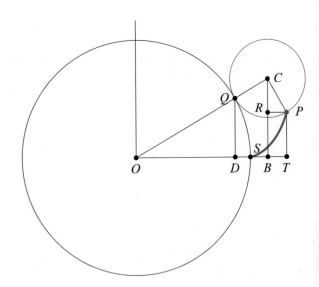

Particular cases of the epicycloid are $a = b$ when a *cardioid* is obtained and $a = 2b$ when a *nephroid* is obtained.

10.1.3.1 Cardioid

Geometrically, a cardioid is an epicycloid with cycles of equal radius. Therefore, the parametric equations of a cardioid are

```
Clear[x,y,a];
x[t_,p_:1]:=2a Cos[t]-p a Cos[2t];
y[t_,p_:1]:=2a Sin[t]-p a Sin[2t];
ParametricPlot[{x[t],y[t]}/.{a→3},{t,0,#},Ticks→{{-8,4.5},{-5,8}},
  ImageSize→{150,150}]&/@(π{1/2,1,3/2,2})
```

See Fig. 10.10.

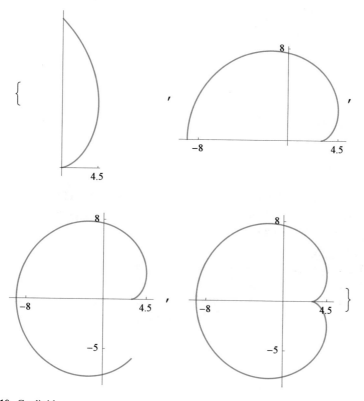

Fig. 10.10 Cardioid

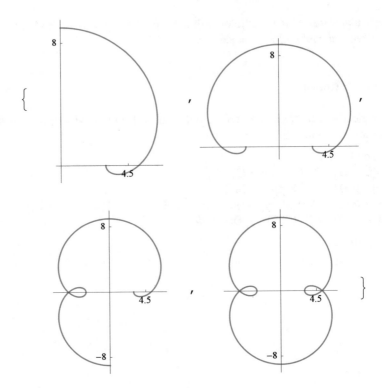

Fig. 10.11 Nephroid

10.1.3.2 Nephroid

As we mentioned earlier, geometrically a nephroid is an epicycloid with the radius
of the rolling circle equal to half of the radius of the fixed cycles. Therefore, the
parametric equations of a nephroid are

```
Clear[x,y,b]
x[t_,p_:1]:=2b Cos[t]-p b Cos[3t]
y[t_,p_:1]:=2b Sin[t]-p b Sin[3t]
ParametricPlot[{x[t],y[t]}/.{b→3},{t,0,#},Ticks→{{-8,4.5},{-8,8}},
ImageSize→{150,150}]&/@(π{1/2,1,3/2,2})
```

See Fig. 10.11.

10.1.3.3 Nephroid of Freeth

The polar equation of the *Freeth's nephroid* is $r = a(1 + 2\sin(\theta/2))$, [19, 6.8,
pp.175–178] or according to http://www-groups.dcs.st-and.ac.uk/~history/Curves/
Freeths.html. Below we use its parametric equations.

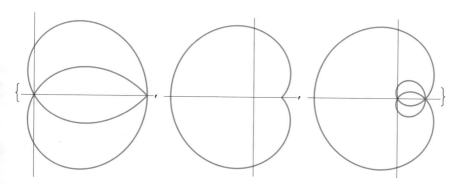

Fig. 10.12 Nephroid of Freeth

```
ParametricPlot[(1+2 Sin[t/2]){Cos[t],Sin[t]},#,Ticks→None,
  ImageSize→150]&/@{{t,-2π,0},{t,0,2π},{t,-2π,2π}}
```
See Fig. 10.12.

10.1.4 Hypocycloid

A *hypocycloid* is the curve described by a point P fixed on the circumference of the circle of radius b which rolls on the inside of the circle of a radius.

Its parametric equations are of the form

```
Clear[x,y,a,b]
x[t_]:=(a-b)Cos[t]+b Cos[(a/b-1)t]
y[t_]:=(a-b)Sin[t]-b Sin[(a/b-1)t]
```

where a and b are parameters. For $a = 8$ and $b = 5$, we have the next graphs.

```
ParametricPlot[{x[t],y[t]}/.{a→8,b→5},{t,0,#},ImageSize→150,
  Ticks→{{-15,15},{-15,15}}]&/@(π{2,4,5,10})
```
See Fig. 10.13.

The above given parametric equations of the hypocycloid are sometimes given under the form

```
xx[t_]:=(a-b)Cos[t]+b Cos[t(k-1)]
yy[t_]:=(a-b)Sin[t]-b Sin[t(k-1)]
b:=a/k;   (* k is a parameter *)
```

Changing the parameters a and k, the resulting figures change drastically.

```
With[{a=1.},
ParametricPlot[{(a-a/#)Cos[t]+(a/#)Cos[t(#-1)],
  (a-a/#)Sin[t]-(a/#)Sin[t(#-1)]},{t,0,10π},ImageSize→150]&/@{.25,.33,
  .5,1.3,1.4,1.6,1.8}]
```
See Fig. 10.14.

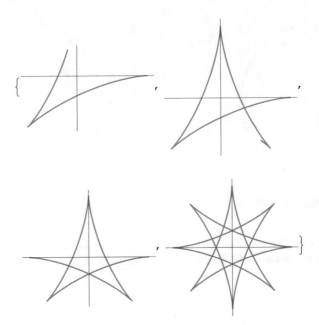

Fig. 10.13 Hypocycloid, 1

The visualization of the geometric definition of the hypocycloid follows.

```
hypocycloid[tend_]:=
Module[{a=4,b=1,x,y,p,xx,gridlines},
x[t_,p_]:=b(a/b-1)Cos[t]+b p Cos[(a/b-1)t];
y[t_,p_]:=b(a/b-1)Sin[t]-b p Sin[(a/b-1)t];
gridlines=Table[{xx,GrayLevel[0.5]},{xx,-a,a,a/5}];
ParametricPlot[{x[t,1],y[t,1]},{t,0,tend},PlotStyle→{Red,Thick},
Ticks→None,ImageSize→225,Epilog→{Thick,Blue,Circle[{0,0},a],
Black,Circle[{x[tend,0],y[tend,0]},b],Red,PointSize[Large],
Point[{x[tend,1],y[tend,1]}],Black,Line[{{x[tend,0],y[tend,0]},
{x[tend,1],y[tend,1]}}]}],GridLines→{gridlines,gridlines},
PlotRange→(1+1/20){-a,a}]]

Manipulate[Quiet@hypocycloid[tend],{tend,.0001,2π,
Appearance→"Labeled"},SaveDefinitions→True]
```

See Fig. 10.15.

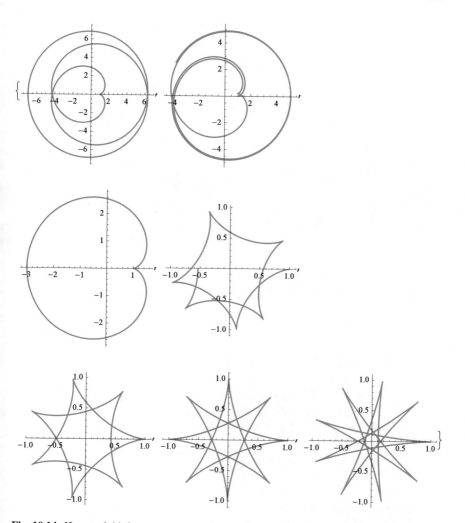

Fig. 10.14 Hypocloid, 2

The proof of the parametric equations of the hypocloid is given here.

```
Clear[a,b,x,y]
With[{a=5,b=1,tend=π/6},
x[t_,p_:1]:=b(a/b-1)Cos[t]+p b Cos[b(a/b-1)t];
y[t_,p_:1]:=b(a/b-1)Sin[t]-p b Sin[b(a/b-1)t];
Show[ParametricPlot[{x[t],y[t]},{t,0,tend},Axes→None,
PlotStyle→{Red,Thick}],
Graphics[{Circle[{0,0},a],{Blue,Circle[{x[tend,0],y[tend,0]},b]},
```

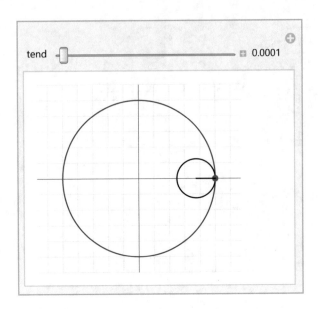

Fig. 10.15 Hypocycloid, 3

Line[{{{0,0},a{Cos[tend],Sin[tend]}},
{{x[tend,0],y[tend,0]},{x[tend,0],0}},
{{x[tend,0],y[tend,0]},{x[tend],y[tend]}},
{{x[tend],y[tend]},{x[tend],0}},{{0,0},{a,0}},
{{x[tend,0],y[tend]},{x[tend],y[tend]}}}],
{Text[Style[O,Italic,12],{0,-.4}],
Text[Style[P,Italic,12],{x[tend]-.33,y[tend]-.25}],
Text[Style[C,Italic,12],{x[tend,0],y[tend,0]+.3}],
Text[Style[T,Italic,12],{x[tend],-.4}],
Text[Style[B,Italic,12],{x[tend,0],-.4}],
Text[Style[S,Italic,12],{a+.35,.15}],
Text[Style[R,Italic,12],{x[tend,0]+.28,y[tend]+.2}],
Text[Style[Q,Italic,12],{a Cos[tend]+.35,a Sin[tend]+.1}]},
{PointSize[Medium],Point[{{0,0},a{Cos[tend],Sin[tend]},
{x[tend,0],y[tend,0]},{x[tend,0],0},{x[tend],0},
{x[tend,0],y[tend]},{a,0}}],Red,Point[{x[tend],y[tend]}]}}],
PlotRange→All,ImageSize→300]]

See Fig. 10.16.

Data: $OQ = OS = a$, $CQ = CP = b$, angle(SOQ)$= t$, arc(SQ)$= at$, arc(QP),
$OB = (a - b) \cos t$, $CB = (a - b) \sin t$, $Q = a(\sin t, \cos t)$, angle(QCP)$= at/b$,
angle(OCB)$= \pi/2 - t$, angle(BCP)$= at/b - t - \pi/2$. □

Fig. 10.16 Hypocycloid, 4

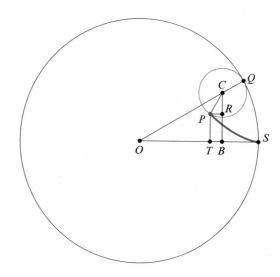

Fig. 10.17 Deltoid

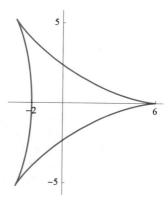

10.1.4.1   Deltoid

A *deltoid* is a particular hypocycloid with $a/b = 3$. We show how it looks like.

```
Clear[x,y,a,b]
x[t_,p_:1]:=(a-b)Cos[t]+p b Cos[(a/b-1)t]
y[t_,p_:1]:=(a-b)Sin[t]-p b Sin[(a/b-1)t]
ParametricPlot[{x[t],y[t]}/.{a→6,b→2},{t,0,2π},
Ticks→{{-2,6},{-5,5}},PlotStyle→Red,ImageSize→150]
```

See Fig. 10.17.

10.1.5   *Maria Agnesi's Curve*

Maria Agnesi's curve is obtained starting with a fixed circle, and a point O on the circle is chosen. For any other point A on the circle, the secant line OA is drawn. The point M is diametrically opposite to O. The line OA intersects the tangent of M at the point N. The line parallel to OM through N and the line perpendicular to OM through A intersect at P. As the point A varies, the path of P is the Maria Agnesi's curve. We consider $O = (0,0)$, $M = (0, 2a)$, and a circle of diameter OM of equation $x^2 + (y - a)^2 = a^2$. Let A be an arbitrary point on this circle. Then $A = a(\sin(2t), 1 - \cos(2a))$, $N = a(2c\tan(t), 2)$, and $P = a(2c\tan(t), 1 - \cos(2t))$. Thus the parametric equations of the curve are $x(t) = 2a \cot t$, $y(t) = a(1 - \cos(2t))$, $a > 0$.

```
With[{a=6,eps=.3},
ParametricPlot[{2a Cot[t], a(1-Cos[2t])},{t,eps,π-eps},
PlotStyle→Red,Ticks→{{-30,30},{a,2a}},Prolog→{{Dashing[{.01}],
Blue,Circle[{0,a},a]}},ImageSize→500]]
```

See Fig. 10.18.

We can generate Maria Agnesi's curve with the next code. The circle is $x^2 + (y - a)^2 = a^2$, where a is a real positive constant. We consider the parameter m varying on the interval $[-\text{mend}, +\text{mend}]$, $(\text{mend} > 0)$.

```
Animate[
With[{a=4,mend=15},
Show[ParametricPlot[{mm,8 a³/(4 a²+mm²)},
{mm,-mend-0.00001,m},Ticks→{{-mend,mend},{a}},
PlotStyle→Red,AxesOrigin→{0,0}], Graphics[{Blue,Circle[{0,a},a],
Line[{{{-mend,2a},{mend,2a}},{{0,0},{m,2a}},
{{4a²m/(4a²+m²),8a³/(4a²+m²)},{m,8a³/(4a²+m²)}},
{{m,0},{m,2a}}}],
PointSize[.015],Black,
Point[{{0,0},{m,2a},{m,0},{4a²m/(4a²+m²),8a³/(4a²+m²)}}],
Red,Point[{{m,8a³/(4a²+m²)}}]],
Text[Style[M,Italic,12],{0,2.2a}],Text[Style[" 0",Italic,12],{0,-1.2}],
Text[Style[N,Italic,12],{m,2.2a}],
Text[Style[P,Italic,12],{m+a/4,8a³/(4a²+m²)-a/5}],}],
PlotRange→All]],{m,-15,15},
AnimationRunning→False,AnimationRate→.3]
```

See Fig. 10.19.

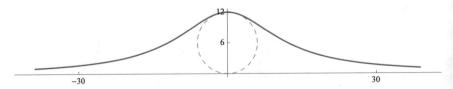

Fig. 10.18 Maria Agnesi's curve

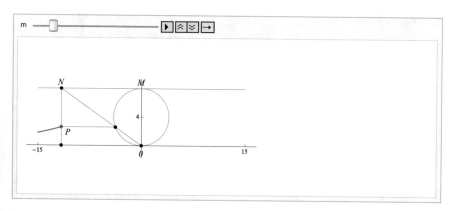

Fig. 10.19 Maria Agnesi's curve, animation

10.1.6   Cassini Ovals

The *Cassini ovals* are described by a point such that the product of its distances from two fixed points (called foci) a distance $2a$ apart is a constant b^2. Immediately the Cartesian equation follows, that is,

$$\left((x-a)^2 + y^2\right)\left((x+a)^2 + y^2\right) = b^2, \quad a, b > 0.$$

The polar equation follows immediately

$$r = a\sqrt{\cos(2t) \pm \sqrt{(b/a)^4 - \sin^2(2t)}}.$$

We note that generally there are two arcs that form the ovals.

```
bmax=2;
cassini[b_]:=Module[{a=1},
Show[PolarPlot[Sqrt[a²(Cos[2t]+Sqrt[(b/a)⁴-Sin[2t]²])]],{t,0,2π},
PlotStyle→Red],
PolarPlot[Sqrt[a²(Cos[2t]-Sqrt[(b/a)⁴-Sin[2t]²])]],{t,0,2π},
PlotStyle→Blue],Ticks→{Automatic,None},
ImageSize→225,PlotRange→All]];

Manipulate[Quiet@cassini[b],
    {{b,1},0,bmax,0.001,Appearance→"Labeled"},SaveDefinitions→True]
```

See Fig. 10.20.

A Cassini oval with $a = b$ is said to be a *Bernoulli lemniscata*.

Fig. 10.20 Cassini ovals

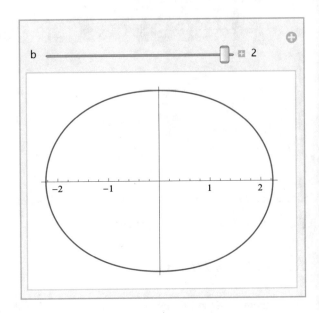

10.1.7 Orthoptics

In plane geometry, an *orthoptic* is the set of points for which two tangents of a given curve meet at a right angle.

Example It is well-known that the orthoptics of

(i) an ellipse $\frac{x^2}{a^2} + \frac{y^2}{b^2} = 1$ is the director (Monge) circle $x^2 + y^2 = a^2 + b^2$, $a > 0, b > 0$;

(ii) a hyperbola $\frac{x^2}{a^2} - \frac{y^2}{b^2} = 1$ is the (Monge) circle $x^2 + y^2 = a^2 - b^2$, $0 < b < a$ (in case of $a \le b$ there are no orthogonal tangents);

(iii) a parabola $y = x^2 / (2p), p > 0$, is its directrix;

(iv) an astroid $x^{2/3} + y^{2/3} = 1$ is a quadrifolium with the polar equation $r = (1/\sqrt{2})\cos(2t), 0 \le t < 2\pi$.

Below we introduce some codes for these orthoptics.

10.1.7.1 Orthoptic of an Ellipse

We consider the ellipse $\frac{x^2}{3^2} + \frac{y^2}{2^2} = 1$ and exhibit its Monge circle and the two orthogonal tangents from each point in the orthoptic circle.

```
Clear[α,m,x]
```

```
mongeEllipse[α_]:=
Module[{a=3,b=2,c,m,m1,m2,t,eps=0.0001,x,x0,y0,x1,y1},
c=Sqrt[a²+b²];pointA=c{Cos[α],Sin[α]};
```

```
ellipse=ParametricPlot[{a Cos[t],b Sin[t]},{t,0,2π},PlotStyle→Blue];
circleMonge=ParametricPlot[c{Cos[t],Sin[t]},{t,0,2π},PlotStyle→Red];
If[Abs[ArcTan[b/a]-α]<eps,pointB={a,0};pointC={0,b};,
If[Abs[π-ArcTan[b/a]-α]<eps,pointB={0,b};pointC={-a,0};,
If[Abs[π+ArcTan[b/a]-α]<eps,pointB={-a,0};pointC={0,-b};,
If[Abs[2π-ArcTan[b/a]-α]<eps,pointB={0,-b};pointC={a,0};,
msol=Solve[(c²Cos[α]²-a²)m²-(c²Sin[2α])m+c²Sin[α]²-b²==0,m];
{m1,m2}={Re[msol[[1,1,2]]],Re[msol[[2,1,2]]]};
sol1=Solve[x²/a2+(c Sin[α]+m1(x-c Cos[α]))²/b²-1==0,x];
x0=Re[sol1[[1,1,2]]];y0=c Sin[α]+m1(x0-c Cos[α]);
pointB={x0,y0};
sol2=Solve[x²/a²+(c Sin[α]+m2(x-c Cos[α]))²/b²-1==0,x];
x1=Re[sol2[[1,1,2]]];y1=c Sin[α]+m2(x1-c Cos[α]);
pointC={x1,y1};];];];];
Show[ellipse,circleMonge,
Graphics[{{Red,Thickness→0.008,Line[{pointC,pointA,pointB}]},
Line[{{a,b},{-a,b},{-a,-b},{a,-b},{a,b}}]},
{PointSize[.018],Point[{pointA,pointB,pointC}]}}],
PlotRange→All,Ticks→{{-4,-3,3,4},{-4,-2,2,4}},ImageSize→300]]

Manipulate[
Quiet@mongeEllipse[α],
{{α,0,"α"},0,2π,.0001,Appearance→"Labeled"},
SaveDefinitions→True,SynchronousUpdating→False]
```

See Fig. 10.21.

10.1.7.2 Orthoptic of a Hyperbola

We consider the hyperbola $\frac{x^2}{3^2} - \frac{y^2}{2^2} = 1$ and exhibit its Monge circle and the two orthogonal tangents from each point in the orthoptic circle. We emphasize that here it is necessary that $a > b > 0$.

```
Clear[α]

mongeHyperbola[α_]:=
Module[{a=3,b=2,c,m,m1,m2,t,eps=0.01,x,x0,y0,x1,y1,pointA,pointB,
pointC},c=Sqrt[a²-b²];pointA=c{Cos[α],Sin[α]};
hyperbola=ParametricPlot[{{a Cosh[t],b Sinh[t]},
{-a Cosh[t],b Sinh[t]}},{t,-π/2,π/2},PlotStyle→{Blue}];
circleMonge=ParametricPlot[c{Cos[t],Sin[t]},{t,0,2π},
PlotStyle→{Red}];
```

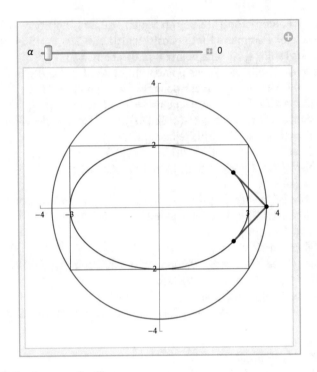

Fig. 10.21   Orthoptic curve of a ellipse

```
If[Abs[ArcTan[b/a]-α]<eps,pointB={a,0};pointC={0,b};,
If[Abs[π-ArcTan[b/a]-α]<eps,pointB={0,b};pointC={-a,0};,
If[Abs[π+ArcTan[b/a]-α]<eps,pointB={-a,0};pointC={0,-b};,
If[Abs[2π-ArcTan[b/a]-α]<eps,pointB={0,-b};pointC={a,0};,
msol=Solve[(c²Cos[α]²-a²)m²-(c²Sin[2α])m+c²Sin[α]²+b²==0,m];
{m1,m2}={Re[msol[[1,1,2]]],Re[msol[[2,1,2]]]};
sol1=Solve[x²/a²-(c Sin[α]+m1(x-c Cos[α]))²/b²-1==0,x];
x0=Re[sol1[[1,1,2]]];y0=c Sin[α]+m1(x0-c Cos[α]);
pointB={x0,y0};
sol2=Solve[x²/a²-(c Sin[α]+m2(x-c Cos[α]))²/b²-1==0,x];
x1=Re[sol2[[1,1,2]]];y1=c Sin[α]+m2(x1-c Cos[α]);
pointC={x1,y1};];];];];
Show[hyperbola,circleMonge,
Graphics[{{Red,Thickness→0.008,Line[{pointC,pointA,pointB}]},
{PointSize[.018],Point[{pointA,pointB,pointC}]}}],
PlotRange→All,Ticks→{{-3,3},{-2,2}},ImageSize→300]]

Manipulate[
Quiet@mongeHyperbola[α],
{{α,0,"α"},-π,π,.0001,Appearance→"Labeled"},
SaveDefinitions→True,SynchronousUpdating→False]
```

See Fig. 10.22.

Fig. 10.22 Orthoptic curve
of a hyperbola

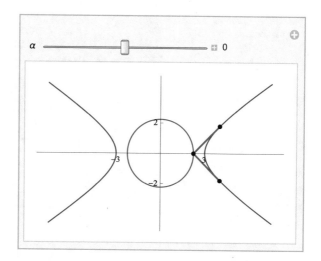

10.1.7.3   Orthoptic of a Parabola

Proposition 10.1 *If two tangents to a parabola are perpendicular to each other,
then they intersect on the directrix. Conversely, two tangents which intersect on the
directrix are perpendicular.*

Proof Let $x^2 = 2py$ be a parabola where p is a positive constant. Suppose that two
tangents contact this parabola at the points $(a, a^2/2p)$ and $(b, b^2/2p)$. The equation
of the first tangent is $y - a^2/2p = (a/p)(x - a)$, whereas the equation of the second
tangent is $y - b^2/2p = (b/p)(x - b)$. From these two equations and the condition of
orthogonality, we get the equations $x = (a + b)/2$ and $y = -p/2$. The last equation
is the equation of the directrix of the given parabola. $\square$

The converse statement is similarly proven.

```
Clear[u]

mongeParabola[u_]:=
Module[{p=2,x},
pointA={u,-p/2};
parabola=ParametricPlot[{x,x²/(2p)},{x,-6.65,6.65},
PlotStyle→{Blue}];
circleMonge=ParametricPlot[{x,-p/2},{x,-6,6},PlotStyle→{Red}];
pointB={u,-p/2};pointA={u-Sqrt[u²+p²],(u-Sqrt[u²+p²])²/(2p)};
pointC={u+Sqrt[u²+p²],(u+Sqrt[u²+p²])²/(2p)};
Show[parabola,circleMonge,
Graphics[{{Red,Thickness→0.008,Line[{pointA,pointB,pointC}]},
{PointSize[.018],Point[{pointA,pointB,pointC}]}}],PlotRange→All,
Ticks→{p{-3/2,3/2},p{-1/2,1/2,1,2,4}},ImageSize→200]]
```

Fig. 10.23 Orthoptic curve
of a parabola

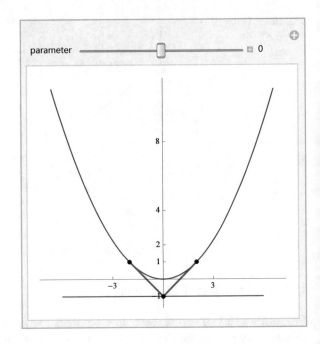

```
Manipulate[
Quiet@mongeParabola[u],
 {{u,0,"parameter"},-3,3,.0001,Appearance→"Labeled"},
 SaveDefinitions→True,SynchronousUpdating→False]
```

See Fig. 10.23.

10.1.7.4   Orthoptic of an Astroid

An astroid can be described by the next parametric representation

$$a(t) = \left(\cos^3 t, \sin^3 t\right), \quad 0 \leq t \leq 2\pi.$$

The tangent vector is $a'(t)$. Then the condition of orthogonality is written
$a'(t) \cdot a'(t + \theta) = 0$, i.e., $\cos(t) = 0$. Immediately one has $\theta = (2k + 1)\pi/2, k \in \mathbb{Z}$;
we set $\theta = \pi/2$. Thus the equations of the orthoptic of an astroid result by solving
the system of equations

$$\begin{cases} y - \sin^3 t = \tan(t)\left(x - \cos^3 t\right) \\ y - \cos^3 t = -\cot(t)\left(x + \sin^3 t\right). \end{cases}$$

Then we get the Cartesian equations of the orthoptic of an astroid is

$$\begin{cases} x(t) = \sin t \cos t(\sin t - \cos t) \\ y(t) = \sin t \cos t(\sin t + \cos t). \end{cases}$$

This curve is said to be a quadrifolium.

```
Clear[a]
mongeAstroid[α_]:=
Module[{a,t},
a[t_]:={Cos[t]³,Sin[t]³};
qfolium[t_]:={Sin[t]Cos[t](Sin[t]-Cos[t]),Sin[t]Cos[t](Sin[t]+Cos[t])};
astroid=ParametricPlot[a[t],{t,0,2π},PlotStyle→Blue];
circleMonge=ParametricPlot[qfolium[t],{t,0,2π},PlotStyle→Red];
Show[astroid,circleMonge,
Graphics[{{Red,Thickness→0.008,Line[{qfolium[α],a[α + π/2]}],
Green,Line[{a[α],qfolium[α]}]},
{PointSize[.018],Point[{a[α],qfolium[α],a[α + π/2]}]}}],
PlotRange→All,Ticks→{{-1,1},{-1,1}},ImageSize→250]]

Manipulate[
Quiet@mongeAstroid[α],
{{α,0,"α"},0,2π,.0001,Appearance→"Labeled"},
SynchronousUpdating→False,SaveDefinitions→True]
```

See Fig. 10.24.

Fig. 10.24 Orthoptic curve
of an astroid

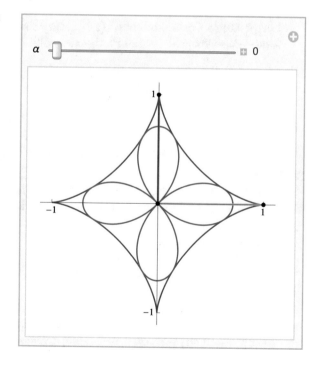

10.1.8   Pedals of Some Curves

The *pedal curve* of a curve and a fixed point is the locus of the orthogonal projection
of the fixed point on the tangent lines of the curve.

Our first example is the pedal curve of a circle. This pedal curve is a cardioid. Its
shape depends on the distance of the point to the center of the circle.

```
Clear[a,b,α]
pedalofCircle[a_,b_,α_]:=   (* a is the radius of the circle; we suppose that
it is centered in the origin of the axes *)
(* b is the distance of the fixed point from the center of the circle *)
Module[{x,y,t,sol},
sol=Solve[a Cos[t]x+a Sin[t]y==a²&&Cos[t] y-Sin[t] x==-b Sin[t],{x,y}];
sol=Flatten[FullSimplify[sol]];
Show[ParametricPlot[{x,y}/.sol,{t,0,2π},PlotStyle→Green],
Graphics[{{Black,Circle[{0,0},a],
PointSize[.02],Point[{{0,0},{b,0},a{Cos[α],Sin[α]}}]},
{Blue,Thickness[.007],
Line[{{a Cos[α],a Sin[α]},{a Cos[α]+b Sin[α]²,(a-b Cos[α])Sin[α]}}]},
{Magenta,PointSize[.03],
Point[{a Cos[α]+b Sin[α]²,(a-b Cos[α])Sin[α]}]},
{Red,Thickness[.007],
Line[{{b,0},{a Cos[α]+b Sin[α]²,(a-b Cos[α])Sin[α]}}]}}],
Axes→False,ImageSize→220,PlotRange→All]]

Manipulate[
Quiet@pedalofCircle[a,b,α],
{{a,1,a},1,4,0.01,Appearance→"Labeled"},
{{b,1.5,b},1,4,0.01,Appearance→"Labeled"},
{{α,0,"α"},0,2π,0.01,Appearance→"Labeled"},
SaveDefinitions→True,SynchronousUpdating→True]
```

See Fig. 10.25.

The second example is the pedal curve of an astroid. The parametric equations
of the astroid are $(a \cos^3 t, b \sin^3 t)$ and the fixed point is (xp, yp).

```
Clear[a,b,xp,yp,α]
pedalofAstroid[a_,b_,xp_,yp_,α_]:=
Module[{x,y,t,sol},   (* a is the horizontal semiaxis of the astroid;
we suppose that is centered in the origin of the axes *)
(* b is the vertical semiaxis of the astroid *)
(* xp is the abscisa of the fixed point and yp is the ordinate of it *)
sol=Flatten[Solve[b Sin[t] x+a Cos[t] y-a b Sin[t] Cos[t]==0&&
a Cos[t] x-b Sin[t] y-a Cos[t] xp +b Sin[t] yp==0,{x,y}]];
```

Fig. 10.25 Pedal curve of a circle

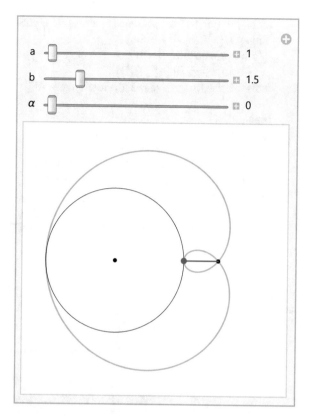

```
Show[ParametricPlot[{{a Cos[t]³,b Sin[t]³},{x,y}/.sol},{t,0,2π},
PlotStyle→{Black,Green}],
Graphics[{{Black,PointSize[.02],
Point[{{0,0},{xp,yp},{a Cos[α]³,b Sin[α]³}}]},
{Blue,Thickness[.007],
Line[{{a Cos[α]³,b Sin[α]³},{-((-a²xp Cos[α]²+a b yp Cos[α]Sin[α]-
a b²Cos[α] Sin[α]²)/(a²Cos[α]²+b²Sin[α]²)),
-((a b xp Cos[α] Sin[α]-a² b Cos[α]² Sin[α]-b²yp Sin[α]²)/
(a² Cos[α]²+b² Sin[α]²))}}]},
{Magenta,PointSize[.03],
Point[{-((-a² xp Cos[α]²+a b yp Cos[α] Sin[α]-a b² Cos[α]Sin[α]²)/
(a² Cos[α]²+b² Sin[α]²)),
-((a b xp Cos[α] Sin[α]-a²b Cos[α]² Sin[α]-b² yp Sin[α]²)/
(a² Cos[α]²+b² Sin[α]²))}]},
{Red,Thickness[.007],
```

```
Line[{{xp,yp},{-((-a²xp Cos[α]²+a b yp Cos[α]Sin[α] -a b²Cos[α]
Sin[α]²)/(a²Cos[α]²+b²Sin[α]²)),
-((a b xpCos[α]Sin[α]-a²b Cos[α]² Sin[α]-b² yp Sin[α]²) /
(a² Cos[α]²+b² Sin[α]²))}}]}}],
Axes→False,ImageSize→220,PlotRange→All]]

Manipulate[
Quiet@pedalofAstroid[a,b,xp,yp,α],
{{a,1,"a"},1,3,0.01,Appearance→"Labeled"},
{{b,1,"b"},1,3,0.01,Appearance→"Labeled"},
{{xp,0,"xp"},0,6,0.01,Appearance→"Labeled"},
{{yp,0,"yp"},0,6,0.01,Appearance→"Labeled"},
{{α,0,"α"},0,2π,0.01,Appearance→"Labeled"},
SaveDefinitions→True,SynchronousUpdating→True]
```

See Fig. 10.26.

Fig. 10.26 Pedal curve of an
astroid

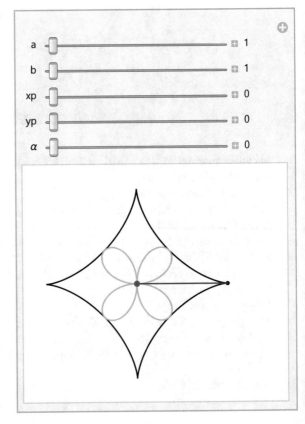

10.2   Attractors

10.2.1   Hénon Attractor

The *Hénon map* is a discrete-time dynamical system, [62, 12.2]. It is one of the most studied examples of dynamical systems that exhibit a chaotic behavior. The Hénon map takes a point (x_n, y_n) in the plane and maps it to a new point according to the recurrences

$$\begin{cases} x_{n+1} = 1 - ax_n^2 + y_n, \\ y_{n+1} = bx_n. \end{cases}$$

The map depends on two real parameters, a and b, which for the classical Hénon map have values of $a = 1.4$ and $b = 0.3$. For the classical values, the Hénon map is chaotic. For other values of a and b, the map may be chaotic, intermittent, or converge to a periodic orbit. Below we show the behavior of the Hénon map for $a = 1.4$ and $b = 0.3$.

```
For[n=2000;x=1;y=1;a=1.4;b=0.3;points={{x,y}};
k=0,k≤n,k++,u=1+y-a x²;v=b x;
AppendTo[points,{u,v}];x=u;y=v]
Show[ListPlot[points,PlotStyle→{Red,PointSize[0.005]},
Axes→False,ImageSize→250],Graphics[{{PointSize[0.012],
Green,Point[First[points]],
PointSize[0.02],Blue,Point[Last[points]]},
Text[Style["initial point",Italic,Darker@Green,12],First[points]+{0,-.1}],
Text[Style["final point",Italic,Blue,12],Last[points]+{-.10,.15}]}]]
```

See Fig. 10.27.
  We study if the Hénon system has an equilibrium point.

```
Clear[x,y]
With[{a=1.4,b=0.3},
Solve[x==1+y-a x²&&y==b x,{x,y},Reals]]
```

Fig. 10.27 The Hénon
attractor

initial point

final point

Solve::ratnz: Solve was unable to solve the system with inexact
coefficients. The answer obtained by solving a corresponding exact system
and numericizing the result. ≫
$\{\{x \to -1.13135, y \to -0.339406\}, \{x \to 0.631354, y \to 0.189406\}\}$

Thus, there are two pairs of real equilibrium points. We test that a point of these
is indeed an equilibrium point.

```
Clear[x,y]
With[{a=1.4,b=0.3},
s=Solve[x==1+y-a x²&&y==b x,{x,y},Reals];
x=x/.s[[1]];y=y/.s[[1]];points={{x,y}};   (* Initial point *)
For[n=2000;k=0,k≤n,k++,u=1+y-a x²;v=b x;
AppendTo[points,{u,v}];x=u;y=v]
Show[ListPlot[points,PlotStyle→{Red,PointSize[0.005]},
Axes→False,ImageSize→200],
Graphics[{{PointSize[0.012],Green,Point[First[points]],
PointSize[0.02],Blue,Point[Last[points]]},
Text[Style["initial point",Italic,Darker@Green,12],First[points]+{0,-.1}],
Text[Style["final point",Italic,Blue,12],Last[points]+{-.10,.15}]}]]]
```
Solve::ratnz: Solve was unable to solve the system with inexact
coefficients. The answer obtained by solving a corresponding exact system
and numericizing the result. ≫

See Fig. 10.28.

It is interesting to consider the Hénon attractors in $\mathbb{R}^3$. Then we get that

```
For[n=2000;x=1;y=1;a=1.4;b=0.3;points={{0,x,y}};
k=0,k≤n,k++,u=1+y-a x²;v=b x;
AppendTo[points,{k,u,v}];x=u;y=v]
Show[ListPointPlot3D[points,ColorFunction→Function[{x,y,z},
Hue[x]]],
Graphics3D[{{PointSize[0.015],Green,Point[First[points]],
PointSize[0.025],Hue[n],Point[Last[points]]},
Text[Style["initial point",Italic,Darker@Green,12],
First[points]+{0,0,-.2}],
Text[Style["final point",Italic,Hue[n],12],Last[points]+{0,.5,-.2}]}]]]
```

See Fig. 10.29.

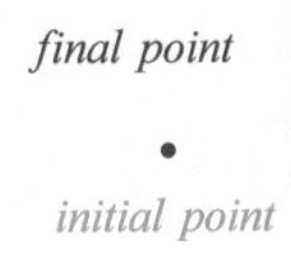

Fig. 10.28 Initial and final
points of a Hénon attractor

final point

initial point

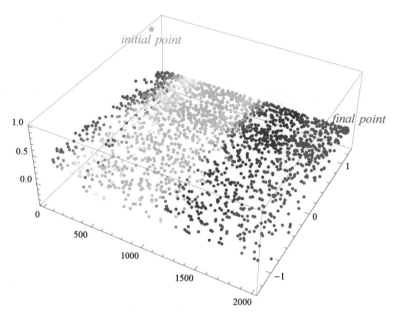

Fig. 10.29 3D Hénon attractor

10.2.2 Lorenz Attractor

The Lorenz attractor was found by Ed. N. Lorenz around 1963, [62, 9.4]. It was derived from a simplified model of convection in the Earth's atmosphere. The system is most commonly expressed as three coupled nonlinear ordinary differential equations.

$$\begin{cases} \frac{dx}{dt} = a(y-x), \\ \frac{dy}{dt} = x(b-z) - y, \\ \frac{dz}{dt} = xy - cz. \end{cases}$$

One commonly used set of constants is $a = 10$, $b = 28$, $c = 8/3$. A code using the Euler's method of integration of this case follows.

```
For[n=300;x=0.0001;y=.0001;z=.0001;a=10;b=28;c=8/3;h=0.015;
point1={x,y,z};points={point1};
k=0,k<n,k++,
u=x+h a(y-x);v=y+h(x(b-z)-y);w=z+h(x y-c z);
AppendTo[points,{u,v,w}];
x=u;y=v;z=w]
point2={x,y,z};
```

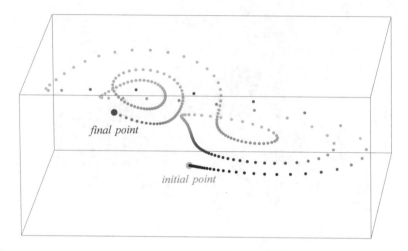

Fig. 10.30 Discret Lorenz attractor

```
Show[
ListPointPlot3D[points,ColorFunction→"Rainbow",
PlotStyle→PointSize[0.008],Axes→False],
Graphics3D[{{PointSize[0.018],Green,Point[point1],Red,Point[point2]},
Text[Style["initial point",Italic,Darker@Green,12],point1+{0,0,-7}],
Text[Style["final point",Italic,Red,12],point2+{0,0,-9}]}],
ViewPoint→{25,2,6},ImageSize→400]
```

See Fig. 10.30.

Here we use the built-in function NDSolve and consider the smooth case of the Lorenz equations. In this case a good source is
http://reference.wolfram.com/language/example/VisualizeTheLorenzAttractor.
html. Then we color the trajectory in two ways.

```
Clear[x,y,z]
Module[{a=10,b=28,c=8/3,tend=100,s,t},
s=NDSolve[{x'[t]==a(y[t]-x[t]),y'[t]==-x[t] z[t]+b x[t]-y[t],
z'[t]==x[t] y[t]-c z[t],x[0]==y[0]==z[0]==0.0001},{x,y,z},{t,0,tend},
MaxSteps→ ∞];
{ParametricPlot3D[Evaluate[{x[t],y[t],z[t]}/.s],{t,0,tend},
PlotPoints→1000,PlotStyle→Directive[Thick,RGBColor[.8,0,0]],
Boxed→False,Axes→False,ImageSize→275,
ColorFunction→(ColorData["SolarColors",#4]&)],
ParametricPlot3D[Evaluate[{x[t],y[t],z[t]}/.s],{t,0,tend},
PlotPoints→1000,Boxed→False,Axes→False,ImageSize→275,
ColorFunction→(ColorData["Rainbow"][#4]&)]}
]
```

The previous two pictures are all static. Below we consider the constants a, b, and c parameters. It is also interesting to us the case when the final instant, here denoted p, is a parameter. This case is interesting because one can see the appearance and evolution of the chaotic movement.

```
Clear[a,b,c,p,x,y,z,xsol,ysol,zsol]
lorenzTrajectory[a_,b_,c_,p_]:=
Module[{x,y,z,xsol,ysol,zsol},
{xsol,ysol,zsol}=NDSolveValue[{x'[t]==a(y[t]-x[t]),y'[t]==x[t](b-z[t])-y[t],
z'[t]==x[t] y[t]-c z[t],x[0]==0.0001,y[0]==0.0001,z[0]==0.0001},{x,y,z},
{t,0,p}];
Show[ParametricPlot3D[{xsol[t],ysol[t],zsol[t]},{t,0,p},PlotRange→All,
PlotPoints→250,ImageSize→250,Boxed→False,Axes→False,
ColorFunction→(ColorData["Rainbow"][#4]&)],
Graphics3D[{{PointSize[0.018],Green,Point[{xsol[0],ysol[0],zsol[0]}]},
Text[Style["initial point",Italic,Darker@Green,12],{xsol[0],ysol[0],
zsol[0]}+{0,0,-6}]}]]]

Manipulate[
Quiet@lorenzTrajectory[a,b,c,p],
{{a,10},9,11,.01,Appearance→"Labeled"},{{b,28},6,40,.01,
Appearance→"Labeled"},
{{c,8/3},1.5,23,.01,Appearance→"Labeled"},{{p,18},0.001,80,.01,
Appearance→"Labeled"},
SaveDefinitions→True,SynchronousUpdating→False]
```

See Fig. 10.31.

10.3 Limit Cycles and Hopf Bifurcation

A *limit cycle* is a planar singular periodic solution of a nonlinear and autonomous system of ordinary differential equations of the form

$$\begin{cases} x' = f(x,y), \\ y' = g(x,y), \end{cases}$$

[52]. We assume that each initial value problem has a solution which is continuable

$$\begin{cases} x = x(t), \\ y = y(t). \end{cases}$$

Fig. 10.31 Continuous
Lorenz attractor

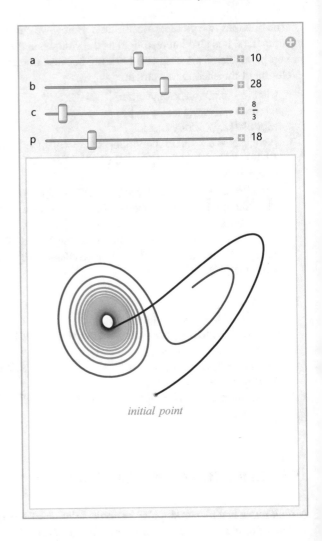

10.3.1 Van Der Pol Equation

The *Van der Pol differential equation* is a second-order nonlinear ordinary differential equation with nonlinear damping:

$$x''(t) - \mu \left(1 - x^2(t)\right) x'(t) + x(t) = 0,$$

where x is the position coordinate depending on the time t and μ is a scalar parameter indicating the nonlinearity and the strength of the damping, [62, pp. 198–199]. This equation can be written under the form of a planar system of differential equations of the first order

$$\begin{cases} x' = y, \\ y' = \mu \left(1 - x^2\right) x' - x. \end{cases}$$

Below we show the existence of a limit cycle for $\mu \in [0, 5]$. The green closed trajectory is the singular periodic orbit. The blue trajectory always remains outside the periodic orbit, whereas the red orbit always remains inside the periodic orbit. These two nonperiodic orbits converge to the periodic one.

```
Clear[μ]

vanderPol[μ_]:= Module[{a1=a3=2,a2=1,b1=3,b2=1,b3=0,x,y,t,sol1,
sol2,sol3,p1,p2,p3,system,point1,point2,point3},
(* Outer trajectory *)
point1={a1,b1};
system={x'[t]==y[t],y'[t]==μ(1-x[t]²)y[t]-x[t],{x[0],y[0]}==point1};
sol1=NDSolve[system,{x,y},{t,100}];
p1=ParametricPlot[Evaluate[{x[t],y[t]}/.sol1],{t,0,100},
PlotStyle→Blue,PlotPoints→200];
freqouter=Plot[x[t]/.sol1,{t,0,100},PlotStyle→Blue,PlotPoints→100,
Ticks→{{50,100},{-3,-2,-1,1,2,3}}];
(* Inner trajectory *)
point2={a2,b2};
system={x'[t]==y[t],y'[t]==μ(1-x[t]²)y[t]-x[t],{x[0],y[0]}==point2};
sol2=NDSolve[system,{x,y},{t,100}];
p2=ParametricPlot[Evaluate[{x[t],y[t]}/.sol2],{t,0,100},PlotStyle→Red,
PlotPoints→300];
freqinner=Plot[x[t]/.sol2,{t,0,100},PlotStyle→Red,PlotPoints→100,
Ticks→{{50,100},{-2,-1,1,2}}];
(* Periodic trajectory; limit cycle *)
point3={a3,b3};
system={x'[t]==y[t],y'[t]==μ(1-x[t]²)y[t]-x[t],{x[0],y[0]}==point3};
sol3=NDSolve[system,{x,y},{t,100}];
p3=ParametricPlot[Evaluate[{x[t],y[t]}/.sol3],{t,0,100},
PlotStyle→Green,PlotPoints→200];
freq=Plot[x[t]/.sol3,{t,0,100},PlotStyle→Green,PlotPoints→100,
Ticks→{{50,100},{-2,-1,1,2}}];
{Show[p1,p2,p3,Graphics[{
Text[Style["initial outer point",Italic,Blue,12],point1+{-.3,0.2}],
Text[Style["initial inner point",Italic,Red,12],point2+{-.5,0.2}],
{PointSize[0.03],Blue,Point[point1],Red,Point[point2],PointSize[0.035],
Green,Point[point3]}}],PlotRange→All,Ticks→{{-2,1,2},{-3,-2,2,3}},
ImageMargins→5,ImageSize→300],
Show[freq,PlotRange→{{0,50},Automatic}],
Show[freqouter,PlotRange→{{0,50},{-2,3}}],
Show[freqinner,PlotRange→{{0,50},{-2,2}}]]}]

Manipulate[Quiet@vanderPol[μ],
{{μ,.4},0,3.1,0.01,Appearance→"Labeled"},SaveDefinitions→True,
SynchronousUpdating→False]
```

See Fig. 10.32.

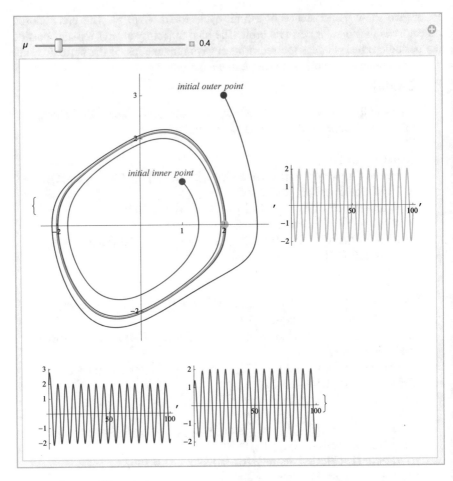

Fig. 10.32 Van der Pol equation

10.3.2   Hopf Bifurcation

As an example of *Hopf bifurcation*, we discuss a planar system of nonlinear differential equations depending on a real parameter μ,   [62, Chap. 8].

$$\begin{cases} x' = \mu - x^2 \\ y' = y. \end{cases}$$

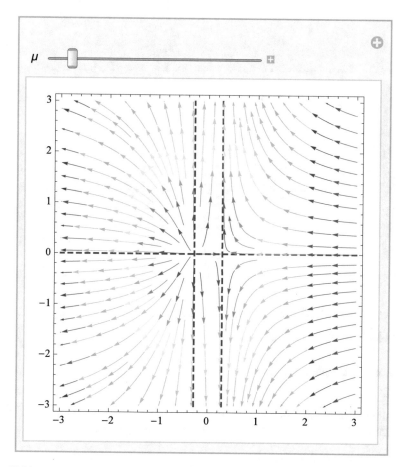

Fig. 10.33   Hopf bifurcation

Clear[μ]

Manipulate[Block[{$PerformanceGoal="Quality"},
Show[{ContourPlot[μ -x^2==0,{x,-3,3},{y,-3,3},Mesh→None,
ContourStyle→{Dashed,Blue}],ContourPlot[y==0,{x,-3,3},{y,-3,3},
ContourStyle→{Dashed,Blue}],
StreamPlot[{μ -x^2,y},{x,-3,3},{y,-3,3},
StreamColorFunction→Hue]},ImageSize→300]],{{μ,0.08},0,1}]

See Fig. 10.33.

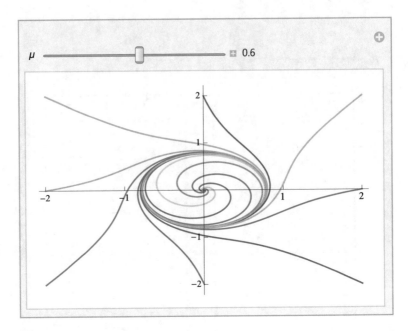

Fig. 10.34 Supercritical Hopf bifurcation

Another example follows.

```
Clear[μ]
```

```
supercriticalHopfbifurcation[μ_]:=
Module[{x,y,u,v,k,h=0.02,nsteps=1000,a=0.001,length,

initpoints,pointss,points},
initpoints=Union[Flatten[Outer[List,{-2,0,2},{-2,0,2}],1],
{{a,0},{0,a},{0,-a},{-a,0}}];
length=Length@initpoints;
pointss={};
Do[points={initpoints[[p]]};{x,y}=Flatten[points];
k=0;
While[k<nsteps,
u=x+h(y-x(x²+y²-μ)); v=y+h(-x-y(x²+y²-μ));
AppendTo[points,{u,v}];x=u;y=v;k++];
AppendTo[pointss,points],{p,length}];
ListLinePlot[pointss,Ticks→{{-2,-1,1,2},{-2,-1,1,2}},
PlotRange→All]]
```

```
Manipulate[
Quiet@supercriticalHopfbifurcation[μ],
{{μ,.6,"μ"},-1,2,0.1,Appearance→"Labeled"},
SynchronousUpdating→False,SaveDefinitions→True]
```

See Fig. 10.34.

Chapter 11
Miscellany in the Euclidean Space

11.1 Some Space Curves

11.1.1 Viviani's Window

Viviani's window is defined as the intersection of the cylinder of radius a and center $(a, 0)$

$$(x - a)^2 + y^2 = a^2, \quad a > 0$$

and the sphere with center $(0, 0, 0)$ and radius $2a$

$$x^2 + y^2 + z^2 = 4a^2, \quad a > 0.$$

In order to see how this curve looks like, we consider the next code:

```
Clear[s,c,x,y]
s=x²+y²+z²-4a²;c=(x-a)²+y²-a²;
Block[{a=1},
ContourPlot3D[{s==0,c==0},{x,-2a,2a},{y,-2a,2a},{z,-2a,2a},
MeshFunctions→{Function[{x,y,z,f},s-c]},MeshStyle→{{Thick,Blue}},
Mesh→{{0}},ContourStyle→Directive[LightGreen,Opacity[0.1],
Specularity[White,30]],Boxed→False,Axes→False,ImageSize→250]]
```

See Fig. 11.1.

It is easy to find its parametric equations

$$x(t) = a(1 + \cos t),$$
$$y(t) = a \sin t,$$
$$z(t) = 2a \sin(t/2), \quad t \in [0, 4\pi].$$

© Springer International Publishing AG 2017
M. Mureşan, *Introduction to Mathematica® with Applications*,
DOI 10.1007/978-3-319-52003-2_11

Fig. 11.1 Generation of the
Viviani's window

Its picture was given above by the red line. We show the Viviani's window
dynamically. The red dot moves on the curve describing it.

```
Clear[a,t]
With[{a=1},
v1=ParametricPlot3D[{a(1+Cos[t]),a Sin[t],2a Sin[t/2]},{t,0,4π},
Boxed→False,Axes→False];
v2=ParametricPlot3D[{{2a{Cos[t],Sin[t],0}},
{2a{Cos[t],0,Sin[t]}},{2a{0,Cos[t],Sin[t]}}},{t,0,2π}];
pointviviani[t_]:=a{1+Cos[t],Sin[t],2Sin[t/2]}];

Animate[Show[v1,v2,
Graphics3D[{PointSize[.04],Red,Point[pointviviani[t]]}],
PlotRange→All,ImageSize→200,ViewPoint→{6a,a/4,a/3}],{t,0,4π},
AnimationRate→.02,AnimationRunning→False,
SaveDefinitions→True]
```

See Fig. 11.2.

11.1.2 The Frenet–Serret Trihedron Along Viviani's Window

Suppose a smooth curve in Euclidean space is given, $r(t)$, with $r'(t) \neq 0$, for all t.
Then we use the following vectors:

$$\text{tangent}(t) = \frac{r'(t)}{\|r'(t)\|},$$

$$\text{normal}(t) = \frac{\text{tangent}'(t)}{\|\text{tangent}'(t)\|},$$

$$\text{binormal}(t) = \text{tangent}(t) \times \text{normal}(t). \quad \text{(cross product)}$$

Fig. 11.2 Dynamic generation of the Viviani's window

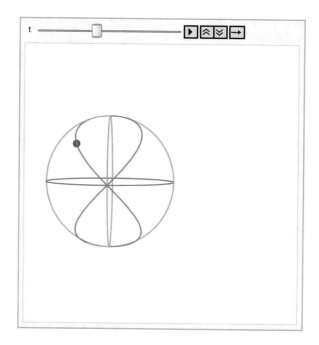

It is known that these vectors are of unitary norm, and each one is orthogonal on the plane determined by the other two. Thus, they define a trihedron called the *Frenet–Serret trihedron*. We do not need to find these three vectors because there is a built-in function FrenetSerretSystem which supplies them. Then we take the Viviani's window and write its corresponding trihedron. The tangent unit vector is colored in green, and the normal unit vector is colored in red, whereas the binormal unit vector is colored in blue.

```
Clear[r,t,u,tangent,normal,binormal]
With[{a=1},
r=a{1+Cos[#],Sin[#],2Sin[#/2]}&;
vectors=Last[FrenetSerretSystem[r[t],t]]//Simplify;
{tangent,normal,binormal}=Map[Arrow[{r[t],r[t]+#}]&,vectors]];

Manipulate[
Show[ParametricPlot3D[r[s],{s,0,4Pi},Boxed→False,Axes→False,
ImageSize→225,PlotStyle→{Thick,Blue},ViewPoint→{8a,a/2,a}],
Graphics3D[{{PointSize[.03],Black,Point[r[t]]},
Thick,Darker[Green],tangent,Red,normal,Blue,binormal}],
PlotRange→2]//Evaluate,{t,0,4Pi,Appearance→{"Labelled"}}]
```

See Fig. 11.3.

Fig. 11.3 Frenet–Serret
trihedron of the Viviani's
window

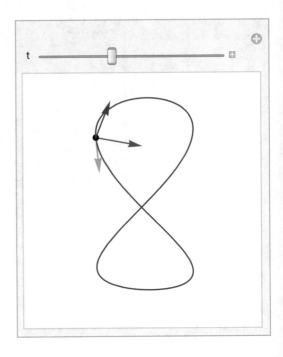

11.1.3   Plane Curves on a Surface

We consider the surface $x = u^2 + u + v$, $y = v^2 + u - v$, and $z = uv$, where u, v are
real parameters. We show that the curves $u = a$ and $v = a$ (a is a constant) on the
surface are planar.

```
(* A necessary and sufficient condition for a curve to lie in a
plane is that its binormal is constant. *)
Clear[a,u,v]
f[u_,v_]:={u²+u+v,v²+u-v,u v}   (* The surface *)
Derivative[2,0][f][u,v]/.v→a   (* A binormal *)
Derivative[0,2][f][u,v]/.u→a   (* The other binormal *)
s=ParametricPlot3D[f[u,v],{u,-5,5},{v,-5,5},Mesh→None,
PlotStyle→Directive[Green,Opacity[0.3],Specularity[White,30]]];
c1=ParametricPlot3D[f[u,2],{u,-5,5},PlotStyle→Blue];
c2=ParametricPlot3D[f[2,v],{v,-5,5},PlotStyle→Red];
Show[s,c1,c2,ImageSize→250,Axes→False,Boxed→False]
{2, 0, 0}
{0, 2, 0}
```

See Fig. 11.4.

Fig. 11.4 Coplanar curves in space

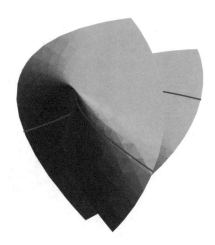

11.2   Some Surfaces in $\mathbb{R}^3$

Along this section we use the tangent, normal, and binormal unit vectors at a point of a space curve to construct interesting surfaces.

11.2.1   Torus

We present a *torus* around the Oz vertical axis.

```
Clear[t,θ,r,u,tube,normal,binormal]
With[{a=1,radius=0.2},
r[t_]:=a{Cos[t],Sin[t],0};
fss[t_]:=FrenetSerretSystem[a{Cos[t],Sin[t],0},t];
normal[u_]:=fss[t][[2,2]]/.t→u;
binormal[u_]:=fss[t][[2,3]]/.t→u;
(* The circle is centered at r[t] in the plane of normal and binormal
unit vectors *)
tube[v_,θ_]:=r[v]+radius(Cos[θ]*normal[v]+Sin[θ]*binormal[v]);
ParametricPlot3D[Evaluate[tube[u,θ]],{u,0,2π},{θ,0,2π},
PlotPoints→{100,12},Ticks→{{-a,0,a},{-a,0,a},{-radius,0,radius}},
MeshStyle→{{Thick,Red}},Mesh→3,
PlotStyle→Directive[Darker[Green],Opacity[0.3],
Specularity[White,40],ImageSize→200]]]
```

See Fig. 11.5.

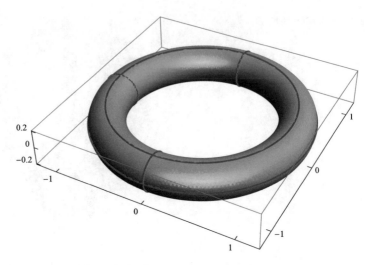

Fig. 11.5 Torus around the vertical axis

11.2.2 A Simple Tube Around Viviani's Window

```
Clear[t,θ,r,u,tube,normal,binormal]
With[{a=1,radius=0.2},
r[t_]:=a{1+Cos[t],Sin[t],2 Sin[t/2]};
fss[t_]:=FrenetSerretSystem[a{1+Cos[t],Sin[t],2Sin[t/2]},t];
normal[u_]:=fss[t][[2,2]]/.t→u;
binormal[u_]:=fss[t][[2,3]]/.t→u;
(* The circle is centered at r[t] in the plane of normal and binormal
unit vectors *)
tube[v_,θ_]:=r[v]+radius(Cos[θ]∗normal[v]+Sin[θ]∗binormal[v]);
ParametricPlot3D[Evaluate[tube[u,θ]],{u,0,4π},{θ,0,2π},
PlotPoints→{100,12},Ticks→{{0,1,2a},{-a,0,a},{-2a,0,2a}},
Mesh→0,PlotStyle→Directive[Darker[Green],Opacity[0.5],
Specularity[White,40]],ImageSize→200]]
```

See Fig. 11.6.

11.2.3 Another Tube Around Viviani's Window

```
Clear[t,θ,r,radius,u,v,tube,normal,binormal]
With[{a=1},
r[v_]:=a{1+Cos[v],Sin[v],2Sin[v/2]};
fss[t_]:=FrenetSerretSystem[a{1+Cos[t],Sin[t],2Sin[t/2]},t];
```

Fig. 11.6 Tube around the
Viviani's window

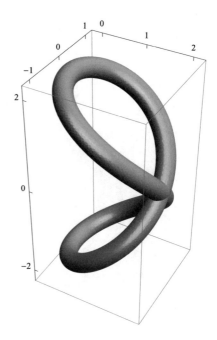

```
normal[u_]:=fss[t][[2,2]]/.t→u;
binormal[u_]:=fss[t][[2,3]]/.t→u;
radius[v_]:=0.2+.04Sin[20v];   (* The radius of the tube *)
(* The circle is centered at r[t] in the plane of normal and binormal
unit vectors *)
Tube[v_,θ_]:=r[v]+radius[v](Cos[θ]*normal[v]+Sin[θ]*binormal[v]);
ParametricPlot3D[Evaluate[tube[u,θ]],{u,0,4 Pi},{θ,0,2 Pi},
PlotPoints→{100,12},Ticks→{{0,1,2},{-1,0,1},{-2,0,2}},
Mesh→None,PlotStyle→Directive[Darker[Green],Opacity[0.5],
Specularity[White,30]],ImageSize→200]]
```

See Fig. 11.7.

11.2.4 Helicoid, Catenoid, and Costa's Minimal Surface

An elliptic helicoid is presented below:

```
With[{a=1,b=3,c=0.5},
ParametricPlot3D[{a u Cos[t],b u Sin[t ],c t},{t,0,2π},{u,0,2},
PlotPoints→80,
```

Fig. 11.7 Another tube
around the Viviani's window

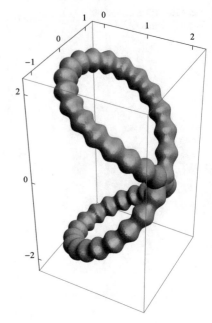

Fig. 11.8 An elliptic helicoid

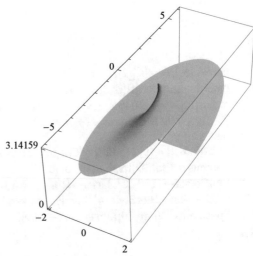

PlotStyle→{Green,Opacity→0.3,Specularity[White,40]},
PlotRange→All,Mesh→None,
Ticks→{{-2a,0,2a},Automatic,{0,2cπ}},ImageSize→225]]

See Fig. 11.8.

Fig. 11.9 Animated helicoid

We animate the above figure in the following way:

```
Animate[
With[{a=1,b=1,c=0.25,v=RotationTransform[-θ,{0,0,1}][{10,0,4}]},
Show[ParametricPlot3D[{a u Cos[t],b u Sin[t],c t},{t,0,3π},{u,0,2},
PlotPoints→50,PlotRange→All,Mesh→None,Boxed→False,
PlotStyle→{Green,Opacity→0.3,Specularity[White,40]},Axes→None,
ImageSize→200],ViewPoint→v]],{θ,0,2π},AnimationRunning→False]
```

See Fig. 11.9.

We now present a catenoid.

```
With[{c=1},
ParametricPlot3D[{c Cosh[u/c]Cos[t],c Cosh[u/c]Sin[t],u},{t,0,2π},
{u,-2,2},
PlotPoints→50,PlotStyle→{Green,Opacity→0.3,
Specularity[White,40]},PlotRange→All,Mesh→None,
ImageSize→250]]
```

See Fig. 11.10.

The circular helicoid can be continuously deformed into a catenoid through a convex combination.

```
Animate[
ParametricPlot3D[{(1-α)u Cos[t]+αCosh[u]Cos[t],
(1-α)u Sin[t]+αCosh[u]Sin[t],t(1-α)+uα},{u,-2,2},{t,0,2 Pi},
PlotRange→All,PlotStyle→{Green,Opacity→0.3,
Specularity[White,40]},Mesh→None,Axes→None,Boxed→False,
```

Fig. 11.10 Catenoid

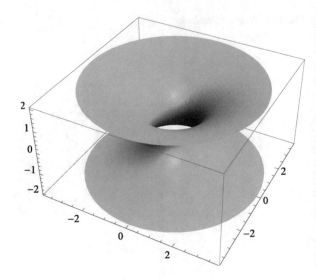

Fig. 11.11 Convex
combination
helicoid-catenoid

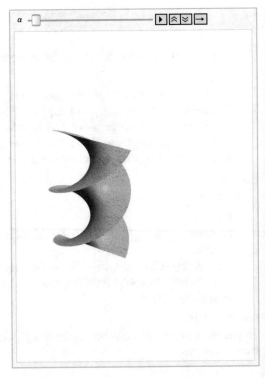

PlotPoints→50,ImageSize→200],{α,0,1},AnimationRate→0.1,
AnimationRunning→False]

See Fig. 11.11.

Fig. 11.12 Costa's minimal
surface

Now we introduce Costa's minimal surface. For reference, see [24]. The code
follows.

```
cw=N[WeierstrassInvariants[{1/2,I/2}],10][[1]];
ew=WeierstrassP[0.5,{cw,0}];
costa[u_,v_]:=
{(1/2)Re[-WeierstrassZeta[u+I v,{cw,0}]+Pi u+Pi²/(2ew)+(Pi/(2ew))
(WeierstrassZeta[u+I v-1/2,{cw,0}]
-WeierstrassZeta[u+I v-I/2,{cw,0}])],
(1/2)Re[-I WeierstrassZeta[u+I v,{cw,0}]+Pi v+Pi²/(2ew)-(Pi I)/(2ew)
(WeierstrassZeta[u+I v-1/2,{cw,0}]
-WeierstrassZeta[u+I v-I/2,{cw,0}])],
(Sqrt[2Pi]/4)Log[Abs[(WeierstrassP[u+I v,{cw,0}]-ew)/
(WeierstrassP[u+I v,{cw,0}]+ew)]]}
costaplot=
With[{eps=10⁻⁶},
ParametricPlot3D[costa[u,v],{u,eps,1-eps},{v,eps,1-eps},
Boxed→False,Axes→False,ViewPoint→{2.9,-1.4,1.5},
PlotStyle→Directive[Green,Opacity→0.3,Specularity[White,40]],
Mesh→3,PlotPoints→80,ImageSize→325]]
```

See Fig. 11.12.

11.3   Some Bodies

11.3.1   The Volume of Two Bodies

In this subsection we find the volumes of two bodies in $\mathbb{R}^3$.

The first body is given by the intersection of a 3D ball with a full cylinder. The
shape of the body is given below:

Fig. 11.13 Intersection of a
ball with a full cylinder

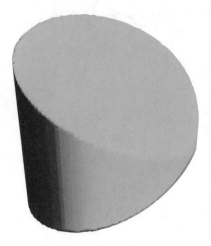

```
Clear[x,y,z]
With[{a=1},
s=x²+y²+z²-4a²;
c=(x-a)²+y²-a²;
reg=RegionPlot3D[s≤0&&c≤0,{x,-2a,2a},{y,-2a,2a},{z,-2a,2a},
PlotStyle→Directive[Green,Opacity[0.3],Specularity[0.5]],
PlotPoints→80,Mesh→None];
Show[reg,PlotRange→All,Boxed→False,Axes→False,
ImageSize→250]]
```

See Fig. 11.13.

The volume of the body is calculated in two ways as follows:

```
Integrate[Boole[s≤0&&c≤0],{x,-2a,2a},{y,-2a,2a},{z,-2a,2a}],
Integrate[If[s≤0&&c≤0,1,0],{x,-2a,2a},{y,-2a,2a},{z,-2a,2a}]
```
$$\left\{ \tfrac{16}{9}(-4 + 3\pi),\ \tfrac{16}{9}(-4 + 3\pi) \right\}$$

The next body is the intersection of a 3D ball with a full double circular
paraboloid.

```
Clear[x,y,z]
With[{a=1},
s=x²+y²+z²-4a²; p=x²+y²-Abs[z];
RegionPlot3D[s≤0&&p≤0,{x,-2a,2a},{y,-2a,2a},{z,-2a,2a},
PlotStyle→Directive[Green,Opacity→0.15,Specularity[White,5]],
PlotPoints→50,Mesh→2,Boxed→False,Axes→False,
ImageSize→250]]
```

See Fig. 11.14.

Fig. 11.14 Double circular
paraboloid

Its volume is found using one of the next codes:

```
FullSimplify[
{Integrate[Boole[s≤0&&p≤0],{x,-2a,2a},{y,-2a,2a},{z,-2a,2a}],
Integrate[If[s≤0&&p≤0,1,0],{x,-2a,2a},{y,-2a,2a},{z,-2a,2a}]}]
```
$$\left\{ \tfrac{1}{6}\left(89 - 17\sqrt{17}\right)\pi, \tfrac{1}{6}\left(89 - 17\sqrt{17}\right)\pi \right\}$$

11.3.2   Reuleaux Tetrahedron

The *Reuleaux tetrahedron* is the intersection of four balls of radius s centered at the vertices of a regular tetrahedron with the side length s; several interesting figures connected to Reuleaux and Meissner tetrahedrons are available at [60, pp. 1194–1195]. The frontier of a ball through each vertex passes through the other three vertices, which also form vertices of the Reuleaux tetrahedron. The Reuleaux tetrahedron has the same face structure as a regular tetrahedron but with curved faces: four vertices and four curved faces, connected by six circular-arc edges. The vertices of the tetrahedron are given below as elements of a list:

$$\text{vertices} = \left\{ \left\{0, 0, -\sqrt{6}\big/4\right\}, \left\{\sqrt{3}\big/3, 0, \sqrt{6}\big/12\right\}, \right.$$
$$\left. \left\{-\sqrt{3}\big/6, 1/2, \sqrt{6}\big/12\right\}, \left\{-\sqrt{3}\big/6, -1/2, \sqrt{6}\big/12\right\} \right\};$$

We verify that the four points are noncoplanar

```
MatrixRank[vertices[[4]]-#&/@Table[vertices[[i]],{i,1,3}]]
3
```

and check that the tetrahedron is regular having the side length 1.

```
Simplify[Norm[{#[[1]]-#[[2]]}]&/@Subsets[vertices,{2}]]
{1, 1, 1, 1, 1, 1}
```

Fig. 11.15 Four spheres

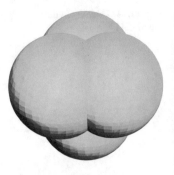

Fig. 11.16 Intersection of
two spheres

Now we can see the four balls.

```
Graphics3D[{Opacity[.5],Green,Specularity[White,1],PlotPoints→50,
Sphere/@vertices},Boxed→False,ImageSize→Small]
```

See Fig. 11.15.

```
Clear[x,y,z]
ineqs=Function[v,v.v][{x,y,z}-#]≤1&/@vertices;
solid=With[{r=.6},
RegionPlot3D[And@@ineqs,{x,-r,r},{y,-r,r},{z,-r,r},
PlotStyle→Directive[Green,Opacity[0.25],Specularity[White,40]],
Mesh→1,PlotPoints→50,Boxed→False,Axes→None,
ImageSize→250]]
```

See Fig. 11.16.

```
tetrahedron=Tetrahedron[vertices];
centroid=RegionCentroid[tetrahedron];
solid=With[{r=.6},
RegionPlot3D[And@@ineqs,{x,-r,r},{y,-r,r},{z,-r,r},
PlotStyle→Directive[Green,Opacity[0.3],Specularity[White,1]],
Mesh→0,PlotPoints→50,Boxed→False,Axes→None,
ImageSize→240] ];
{solidobject=Show[solid,Graphics3D[{Pink,tetrahedron}],
PlotRange→All,ImageSize→240],Rotate[solidobject,π/3,{0,0}]}
```

Fig. 11.17 Tetrahedrons

Fig. 11.18 Animated
tetrahedron

See Fig. 11.17.

```
Animate[
Rotate[solidobject,θ,{0,0}],{θ,0,2π},AnimationRate→0.1,
SaveDefinitions→True,AnimationRunning→False]
```

See Fig. 11.18.

11.4   Local Extrema of Real Functions of Several Real Variables

11.4.1   Unconstrained Local Extrema

11.4.1.1   The First Example

We introduce a code to find the local extrema which is suitable for functions of two variables.

```
Clear[x,y,f]
f[x_,y_]:=y^4-y^3-3x^2 y+x^4
{dfx[x_,y_]=D[f[x,y],x],dfy[x_,y_]=D[f[x,y],y],dfxx[x_,y_]=D[f[x,y],x,x],
dfyy[x_,y_]=D[f[x,y],y,y],dfxy[x_,y_]=D[f[x,y],x,y],
dff[x_,y_]=dfxx[x,y]*dfyy[x,y]-dfxy[x,y]^2};
solns=DeleteDuplicates[{x,y}/.Solve[{dfx[x,y]==0,dfy[x,y]==0},{x,y}]];
If[Length@solns==0,Print[" No stationary point"],
solnsheader={"no.","stationary point"};
Print[Grid[Join[{solnsheader},Table[Prepend[{solns[[k]]},k],
{k,Length@solns}]],Frame→All,FrameStyle→Thin,
Alignment→Right]];
realsolns=Select[solns,FreeQ[#,Complex]&];
If[Length@realsolns==0,Print[" No real stationary point"],
realsolnsheader={"no.","real stationary point"};
Print[Grid[Join[{realsolnsheader},Table[Prepend[{realsolns[[k]]},k],
{k,Length@realsolns}]],Frame→All,FrameStyle→Thin,
Alignment→Right]];
extremum=
Select[realsolns,dff[#[[1]],#[[2]]]>0&&dfxx[#[[1]],#[[2]]]≠0&];
If[Length[extremum]==0,Print[" No point of extremum"],
line={};
Do[If[dfxx[extremum[[k,1]],extremum[[k,2]]]<0,text=" maximum",
text=" minimum"];
line=AppendTo[line,{k,extremum[[k]],text,f@@extremum[[k]]}],
{k,Length@extremum}];
extremumheader={"no.","point of extremum","nature",
"value of the function"};
Print[Grid[Join[{extremumheader},Table[line[[k]],
{k,Length@extremum}]],
Frame→All,FrameStyle→Thin,Alignment→Right]]]]]]
```

Fig. 11.19 The surface under discussion

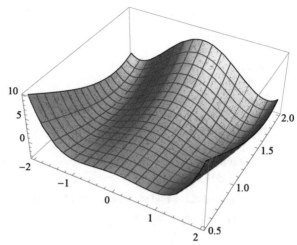

No.	Stationary point
1	$\left\{-\frac{3i}{2\sqrt{2}}, -\frac{3}{4}\right\}$
2	$\left\{\frac{3i}{2\sqrt{2}}, -\frac{3}{4}\right\}$
3	$\{0, 0\}$
4	$\left\{0, \frac{3}{4}\right\}$
5	$\left\{-\frac{3}{2}, \frac{3}{2}\right\}$
6	$\left\{\frac{3}{2}, \frac{3}{2}\right\}$

No.	Real stationary point
1	$\{0, 0\}$
2	$\left\{0, \frac{3}{4}\right\}$
3	$\left\{-\frac{3}{2}, \frac{3}{2}\right\}$
4	$\left\{\frac{3}{2}, \frac{3}{2}\right\}$

No.	Point of extremum	Nature	Value of the function
1	$\left\{-\frac{3}{2}, \frac{3}{2}\right\}$	minimum	$-\frac{27}{8}$
2	$\left\{\frac{3}{2}, \frac{3}{2}\right\}$	minimum	$-\frac{27}{8}$

Now we visualize this function.

```
Plot3D[f[x,y],{x,-2,2},{y,0.5,2},ImageSize→250]
```

See Fig. 11.19.

Example The previous code works properly only if the nature of the points of extremum are sensitive to the first and second order differentials. The situation is similar to the one-dimensional case $f(x) = x^3, x \in \mathbb{R}$, where higher order derivatives are necessary to settle down the nature of the stationary point.

11.4.1.2   The Second Example

The example we discuss below appears in [53, Chap. 7]. The statement of the exercise reads as follows: find the local extreme values of the following function

$$f(x, y, z) = 2x^2 - xy + 2xz - y + y^3 + z^2, \quad x, y, z \in \mathbb{R}.$$

First we approach it by the built-in functions NMinimize and NMaximize.

```
Clear[x,y,z,f]
variables={x,y,z};
f[x_,y_,z_]:=2x²-x y+2x z-y+y³+z²
NMinimize[f[x,y,z],variables]
NMaximize[f[x,y,z],variables]
```

$\{-0.481481, \{x \to 0.333333, y \to 0.666667, z \to -0.333333\}\}$
NMaximize::cvdiv : Failed to converge to a solution. The function may be unbounded. ≫
$\{9.67408528878118 \times 10^{339}, \{x \to 1.4507 \times 10^{9}, y \to 2.13077 \times 10^{113},$
$z \to 9.0809 \times 10^{8}\}\}$

We extract the coordinates of the point of minimum.

```
{x,y,z}/.Last[NMinimize[f[x,y,z],variables]]
```

$\{0.333333, 0.666667, -0.333333\}$

We also extract

```
x/.Last[NMinimize[f[x,y,z],variables]]
y/.Last[NMinimize[f[x,y,z],variables]]
x/.Last[NMinimize[f[x,y,z],variables]]
```

0.333333
0.666667
−0.333333

The next approach is based on the built-in functions Maximize and Minimize.

```
Clear[x,y,z]
Minimize[f[x,y,z],{x,y,z}]
Maximize[f[x,y,z],{x,y,z}]
```

Minimize::natt: The minimum is not attained at any point satisfying the given constraints ≫
$\left\{-\infty, \left\{x \to \frac{33}{10}, y \to -\infty, z \to \frac{8}{5}\right\}\right\}$
Maximize::natt: The maximum is not attained at any point satisfying the given constraints ≫
$\left\{\infty, \left\{x \to -\infty, y \to \frac{33}{10}, z \to \frac{8}{5}\right\}\right\}$

We conclude that this function has neither global maximum point nor global minimum point.

Now we make use of the tools of differential calculus and find the stationary points.

```
Clear[x,y,z]
variables={x,y,z};
f[x_,y_,z_]:=2x²-x y+2x z-y+y³+z²
solns=DeleteDuplicates[
variables/.Solve[Thread[Grad[f[x,y,z],variables]==0],variables]];
If[Length[solns]==0,Print[" No stationary point"],
solnsheader={"no.","stationary point"};
Print[Grid[Join[{solnsheader},Table[Prepend[{solns[[k]]},k],
{k,Length[solns]}]]],Frame→All,FrameStyle→Thin,Alignment→Right]];
realsolns=Select[solns,FreeQ[#,Complex]&];
If[Length[realsolns]==0,Print[" No real stationary point"],
realsolnsheader={"no.","real stationary point"};
Print[Grid[Join[{realsolnsheader},Table[Prepend[{realsolns[[k]]},k],
{k,Length[realsolns]}]]],Frame→All,FrameStyle→Thin,
Alignment→Right]];
seconddiff[x_,y_,z_]:=D[f[x,y,z],{variables,2}];
extremum={};
Do[matrix=seconddiff[x,y,z]/.Thread[variables→realsolns[[j]]];
If[PositiveDefiniteMatrixQ[matrix]||NegativeDefiniteMatrixQ[matrix],
AppendTo[extremum,realsolns[[j]]]],{j,Length@realsolns}];
If[Length[extremum]==0,Print[" No point of extremum"],
line={};
Do[matrix=seconddiff[x,y,z]/.Thread[variables→extremum[[k]]];
If[NegativeDefiniteMatrixQ[matrix],text=" maximum",text=" minimum"];
line=AppendTo[line,{k,extremum[[k]],text,f@@extremum[[k]]}],
{k,Length@extremum}];
extremumheader={"no.","point of extremum","nature",
"value of the function"};
Grid[Join[{extremumheader},Table[line[[k]],{k,Length@extremum}]],
Frame→All,FrameStyle→Thin,Alignment→Right]
]]]
```

No.	Stationary point	No.	Real stationary point
1	$\{-\frac{1}{4},-\frac{1}{2},\frac{1}{4}\}$	1	$\{-\frac{1}{4},-\frac{1}{2},\frac{1}{4}\}$
2	$\{\frac{1}{3},\frac{2}{3},-\frac{1}{3}\}$	2	$\{\frac{1}{3},\frac{2}{3},-\frac{1}{3}\}$

No.	Point of extremum	Nature	Value of the function
1	$\{\frac{1}{3},\frac{2}{3},-\frac{1}{3}\}$	minimum	$-\frac{13}{27}$

11.4.2   Constrained Local Extrema

11.4.2.1   The First Example

Along this subsection we consider the problem of finding the extreme values of the function

$$f[x, y, z] = xyz, \quad \text{subject to} \quad \begin{cases} x^2 + y^2 + z^2 = 1, \\ x + y + z = 1, \quad x, y, z \in \mathbb{R}. \end{cases}$$

In order to see where these results come from, we use the *method of Lagrange multipliers* [53, Chap. 7].

We have three real variables and two constraints. Consequently, we have two multipliers.

```
Clear[x,y,z,λ,μ]
function[x_,y_,z_]:=x y z
variables={x,y,z};
condition1[x_,y_,z_]:=x²+y²+z²-1
condition2[x_,y_,z_]:=x+y+z
multipliers={λ,μ};
allvariables=Flatten[{variables,multipliers}];
```

The function we need is the Lagrangian

```
lagrangian[x_,y_,z_,λ_,μ_]:={function[x,y,z],condition1[x,y,z],
condition2[x,y,z]}.{1,λ,μ}
```

We look for the stationary points.

```
solns=DeleteDuplicates[allvariables/.
Solve[Thread[Grad[lagrangian[x,y,z,λ,μ],allvariables]==0],
allvariables]];
If[Length@solns==0,Print["No stationary point"],
Print[Grid[Join[{{"no.","stationary point"}},
Table[Prepend[{Simplify[solns[[k]]]},k],{k,Length@solns}]],
Frame→All,FrameStyle→Thin,Alignment→Right]];
realsolns=Select[solns,FreeQ[#,Complex]&];
If[Length@realsolns==0,Print["No real stationary point"],
Print[Grid[Join[{{"no.","real stationary point"}},
Table[Prepend[{Simplify[realsolns[[k]]]},k],{k,Length@realsolns}]],
Frame→All,FrameStyle→Thin,Alignment→Right]];
seconddiff[x_,y_,z_,λ_,μ_]=D[lagrangian[x,y,z,λ,μ],{allvariables,2}];
dxyz={dx,dy,dz};
xyzsolns=Table[Take[realsolns[[k]],3],{k,Length@realsolns}];
cond1:=Table[D[condition1[x,y,z],{variables,1}].dxyz/.
Thread[variables→xyzsolns[[j]]],{j,Length@realsolns}];
```

```
cond2:=Table[D[condition2[x,y,z],{variables,1}].dxyz/.
Thread[variables→xyzsolns[[j]]],{j,Length@realsolns}];
soldydz=Table[Flatten[Solve[cond1[[k]]==0&&cond2[[k]]==0,
{dx,dy,dz},Reals]],{k,Length@realsolns}];
soldydz1=dxyz/.soldydz;
dxyzall=Table[Join[soldydz1[[k]],{dλ,dμ}],{k,Length@realsolns}];
lines={};
Do[
matrix=Simplify[
dxyzall[[k]].(seconddiff[x,y,z,λ,μ]/.
Thread[allvariables→realsolns[[k]]]).dxyzall[[k]]];
value=FullSimplify[function[x,y,z]/.Thread[variables→xyzsolns[[k]]]];
point=Simplify[xyzsolns[[k]]];
If[PositiveDefiniteMatrixQ[matrix],
lines=Append[lines,{k,point,matrix,value}],
If[NegativeDefiniteMatrixQ[matrix],
lines=Append[lines,{k,point,matrix,value}],
lines=Append[lines,{k,point,matrix,value}]]],
{k,Length@realsolns}]]]
header={"no.","point of extremum","second order diff.","value of
the function"};
Grid[Join[{header},Table[lines[[k]],{k,Length@realsolns}]],
Frame→All,FrameStyle→Thin,Alignment→Right]
```

No.	Stationary point
1	$\left\{-\sqrt{\frac{2}{3}},\frac{1}{\sqrt{6}},\frac{1}{\sqrt{6}},\frac{1}{2\sqrt{6}},\frac{1}{6}\right\}$
2	$\left\{\sqrt{\frac{2}{3}},-\frac{1}{\sqrt{6}},-\frac{1}{\sqrt{6}},-\frac{1}{2\sqrt{6}},\frac{1}{6}\right\}$
3	$\left\{-\frac{1}{\sqrt{6}},-\frac{1}{\sqrt{6}},\sqrt{\frac{2}{3}},-\frac{1}{2\sqrt{6}},\frac{1}{6}\right\}$
4	$\left\{-\frac{1}{\sqrt{6}},\sqrt{\frac{2}{3}},-\frac{1}{\sqrt{6}},-\frac{1}{2\sqrt{6}},\frac{1}{6}\right\}$
5	$\left\{\frac{1}{\sqrt{6}},-\sqrt{\frac{2}{3}},\frac{1}{\sqrt{6}},\frac{1}{2\sqrt{6}},\frac{1}{6}\right\}$
6	$\left\{\frac{1}{\sqrt{6}},\frac{1}{\sqrt{6}},-\sqrt{\frac{2}{3}},\frac{1}{2\sqrt{6}},\frac{1}{6}\right\}$

No.	Real stationary point
1	$\left\{-\sqrt{\frac{2}{3}},\frac{1}{\sqrt{6}},\frac{1}{\sqrt{6}},\frac{1}{2\sqrt{6}},\frac{1}{6}\right\}$
2	$\left\{\sqrt{\frac{2}{3}},-\frac{1}{\sqrt{6}},-\frac{1}{\sqrt{6}},-\frac{1}{2\sqrt{6}},\frac{1}{6}\right\}$
3	$\left\{-\frac{1}{\sqrt{6}},-\frac{1}{\sqrt{6}},\sqrt{\frac{2}{3}},-\frac{1}{2\sqrt{6}},\frac{1}{6}\right\}$
4	$\left\{-\frac{1}{\sqrt{6}},\sqrt{\frac{2}{3}},-\frac{1}{\sqrt{6}},-\frac{1}{2\sqrt{6}},\frac{1}{6}\right\}$
5	$\left\{\frac{1}{\sqrt{6}},-\sqrt{\frac{2}{3}},\frac{1}{\sqrt{6}},\frac{1}{2\sqrt{6}},\frac{1}{6}\right\}$
6	$\left\{\frac{1}{\sqrt{6}},\frac{1}{\sqrt{6}},-\sqrt{\frac{2}{3}},\frac{1}{2\sqrt{6}},\frac{1}{6}\right\}$

Solve::svars: Equations may not give solutions for all "solve" variables. ≫
Solve::svars: Equations may not give solutions for all "solve" variables. ≫
Solve::svars: Equations may not give solutions for all "solve" variables. ≫
General::stop: Further output of Solve::svars will be suppressed during this
calculation. ≫

No.	Point of extremum	Second order diff.	Value of the function
1	$\left\{-\sqrt{\frac{2}{3}},\frac{1}{\sqrt{6}},\frac{1}{\sqrt{6}}\right\}$	$\sqrt{6}dy^2$	$-\frac{1}{3\sqrt{6}}$
2	$\left\{\sqrt{\frac{2}{3}},-\frac{1}{\sqrt{6}},-\frac{1}{\sqrt{6}}\right\}$	$-\sqrt{6}dy^2$	$\frac{1}{3\sqrt{6}}$
3	$\left\{-\frac{1}{\sqrt{6}},-\frac{1}{\sqrt{6}},\sqrt{\frac{2}{3}}\right\}$	$-\sqrt{6}dx^2$	$\frac{1}{3\sqrt{6}}$
4	$\left\{-\frac{1}{\sqrt{6}},\sqrt{\frac{2}{3}},-\frac{1}{\sqrt{6}}\right\}$	$-\sqrt{6}dx^2$	$\frac{1}{3\sqrt{6}}$
5	$\left\{\frac{1}{\sqrt{6}},-\sqrt{\frac{2}{3}},\frac{1}{\sqrt{6}}\right\}$	$\sqrt{6}dx^2$	$-\frac{1}{3\sqrt{6}}$
6	$\left\{\frac{1}{\sqrt{6}},\frac{1}{\sqrt{6}},-\sqrt{\frac{2}{3}}\right\}$	$\sqrt{6}dx^2$	$-\frac{1}{3\sqrt{6}}$

We conclude that the first, fifth, and sixth points are points of minimum, whereas the other ones are points of maximum. The values of the function on all these points have been already established in the last column.

11.4.2.2   The Second Example

We want to find the points of local extreme of the function

$$f(x,y,z) = 2x^2 + 2y^2 - xy + z^4 - 2z^2, \quad \text{on the compact set}$$

$$K = \left\{(x,y,z)\,\middle|\,x^2 + y^2 + 2z^2 \le 8\right\}, \quad x,y,z \in \mathbb{R}.$$

We have two cases. The cases are exclusive and exhaustive.

Firstly, we find the points of local extreme lying in the interior of the compact set K. Therefore, we repeat the algorithm in the unconstraint local extreme case.

```
Clear[x,y,z,function]
variables={x,y,z};
function[x_,y_,z_]:=2x^2+2y^2-x y+z^4-2z^2;
```

Then we find the stationary points.

```
solns=Select[DeleteDuplicates[
variables/.Solve[Thread[Grad[function[x,y,z],variables]==0],variables]],
#[[1]]^2 + #[[2]]^2 + 2#[[3]]^2 < 8&];
If[Length[solns]==0,Print["No stationary point interior"],
solnsheader={"no.","stationary point interior"};
Print[Grid[Join[{solnsheader},Table[Prepend[{solns[[k]]},k],
{k,Length[solns]}]],Frame→All,FrameStyle→Thin,Alignment→Right]];
realsolns=Select[solns,FreeQ[#,Complex]&];
If[Length[realsolns]==0,Print["No real stationary point interior"],
realsolnsheader={"no.","real stationary point interior"};
```

```
Print[Grid[Join[{realsolnsheader},Table[Prepend[{realsolns[[k]]},k],
{k,Length[realsolns]}]],Frame→All,FrameStyle→Thin,
Alignment→Right]];
seconddiff[x_,y_,z_]:=D[function[x,y,z],{variables,2}];
extremum={};
Do[matrix=seconddiff[x,y,z]/.Thread[variables→realsolns[[j]]];
If[PositiveDefiniteMatrixQ[matrix]||NegativeDefiniteMatrixQ[matrix],
AppendTo[extremum,realsolns[[j]]]],{j,Length@realsolns}];
If[Length[extremum]==0,Print[" No point of extremum interior"],
line={};
Do[matrix=seconddiff[x,y,z]/.Thread[variables→extremum[[k]]];
If[NegativeDefiniteMatrixQ[matrix],text="maximum",text="minimum"];
line=AppendTo[line,{k,extremum[[k]],text,function@@extremum[[k]]}],
{k,Length@extremum}];
extremumheader={" no."," point of extremum"," nature"," value of the
function" };
Grid[Join[{extremumheader},Table[line[[k]],{k,Length@extremum}]],
Frame→All,FrameStyle→Thin,Alignment→Right]
]]]
```

No.	Stationary point interior K
1	$\{0,0,-1\}$
2	$\{0,0,0\}$
3	$\{0,0,1\}$

No.	Real stationary point interior K
1	$\{0,0,-1\}$
2	$\{0,0,0\}$
3	$\{0,0,1\}$

No.	Point of extremum	Nature	Value of the function
1	$\{0,0,-1\}$	minimum	-1
2	$\{0,0,1\}$	minimum	-1

Let us look for the points of extreme located on the boundary of K. Then we apply the method of Lagrange multipliers.

```
Clear[x,y,z,condition,lagrangian,λ,seconddiff]
condition[x_,y_,z_]:=x²+y²+2z²-8;
multiplier={λ};
allvariables=Flatten[{variables,multiplier}];
```

The function we need is

```
lagrangian[x_,y_,z_,λ_]:={function[x,y,z],condition[x,y,z]}.{1,λ}
```

We look for the stationary points

```
solns=DeleteDuplicates[allvariables/.
Solve[Thread[Grad[lagrangian[x,y,z,λ],allvariables]==0],allvariables]];
If[Length@solns==0,Print[" No critical point on the boundary of K"],
Print[Grid[Join[{{" no."," critical point on the boundary of K"}},
Table[Prepend[{Simplify[solns[[k]]]},k],{k,Length@solns}]],
```

```
Frame→All,FrameStyle→Thin,Alignment→Right]];
realsolns=Select[solns,FreeQ[#,Complex]&];
If[Length@realsolns==0,Print["No real critical point on the boundary of
K"],
Print[Grid[Join[{{"no.","real critical point on the boundary
of K"}},Table[Prepend[{Simplify[realsolns[[k]]]},k],
{k,Length@realsolns}]],Frame→All,FrameStyle→Thin,
Alignment→Right]];
seconddiff[x_,y_,z_,λ_]=D[lagrangian[x,y,z,λ],{allvariables,2}];
dxyz={dx,dy,dz};
xyzsolns=Table[Take[realsolns[[k]],3],{k,Length@realsolns}];
cond:=Table[D[condition[x,y,z],{variables,1}].dxyz/.
Thread[variables→xyzsolns[[j]]],{j,Length@realsolns}];
soldydz=Table[Flatten[Solve[cond[[k]]==0,{dx,dy,dz},Reals]],
{k,Length@realsolns}];
soldydz1=dxyz/.soldydz;
dxyzall=Table[Join[soldydz1[[k]],{dλ}],{k,Length@realsolns}];
lines={};
Do[
matrix=Simplify[
dxyzall[[k]].(seconddiff[x,y,z,λ]/.Thread[allvariables→realsolns[[k]]]).
dxyzall[[k]]];
value=FullSimplify[function[x,y,z]/.Thread[variables→xyzsolns[[k]]]];
point=Simplify[xyzsolns[[k]]];
If[PositiveDefiniteMatrixQ[matrix],
lines=Append[lines,{k,point,matrix,value}],
If[NegativeDefiniteMatrixQ[matrix],
lines=Append[lines,{k,point,matrix,value}],
lines=Append[lines,{k,point,matrix,value}]]],{k,Length@realsolns}]
]]
header={"no.","point of extremum","second order diff.","value of f",
"nature"};
Grid[Join[{header},Table[lines[[k]],{k,Length@realsolns}]],
Frame→All,FrameStyle→Thin,Alignment→Right]
```

No.	Critical point on the boundary of K
1	$\{-2, -2, 0, -\frac{3}{2}\}$
2	$\{-2, 2, 0, -\frac{5}{2}\}$
3	$\{0, 0, -2, -3\}$
4	$\{0, 0, 2, -3\}$
5	$\{2, -2, 0, -\frac{5}{2}\}$
6	$\{2, 2, 0, -\frac{3}{2}\}$
7	$\{-\sqrt{\frac{3}{2}}, -\sqrt{\frac{3}{2}}, -\sqrt{\frac{5}{2}}, -\frac{3}{2}\}$
8	$\{-\sqrt{\frac{3}{2}}, -\sqrt{\frac{3}{2}}, \sqrt{\frac{5}{2}}, -\frac{3}{2}\}$
9	$\{\sqrt{\frac{3}{2}}, \sqrt{\frac{3}{2}}, -\sqrt{\frac{5}{2}}, -\frac{3}{2}\}$
10	$\{\sqrt{\frac{3}{2}}, \sqrt{\frac{3}{2}}, \sqrt{\frac{5}{2}}, -\frac{3}{2}\}$
11	$\{-\frac{1}{\sqrt{2}}, \frac{1}{\sqrt{2}}, -\sqrt{\frac{7}{2}}, -\frac{5}{2}\}$
12	$\{-\frac{1}{\sqrt{2}}, \frac{1}{\sqrt{2}}, \sqrt{\frac{7}{2}}, -\frac{5}{2}\}$
13	$\{\frac{1}{\sqrt{2}}, -\frac{1}{\sqrt{2}}, -\sqrt{\frac{7}{2}}, -\frac{5}{2}\}$
14	$\{\frac{1}{\sqrt{2}}, -\frac{1}{\sqrt{2}}, \sqrt{\frac{7}{2}}, -\frac{5}{2}\}$

No.	Real critical point on the boundary of K
1	$\{-2, -2, 0, -\frac{3}{2}\}$
2	$\{-2, 2, 0, -\frac{5}{2}\}$
3	$\{0, 0, -2, -3\}$
4	$\{0, 0, 2, -3\}$
5	$\{2, -2, 0, -\frac{5}{2}\}$
6	$\{2, 2, 0, -\frac{3}{2}\}$
7	$\{-\sqrt{\frac{3}{2}}, -\sqrt{\frac{3}{2}}, -\sqrt{\frac{5}{2}}, -\frac{3}{2}\}$
8	$\{-\sqrt{\frac{3}{2}}, -\sqrt{\frac{3}{2}}, \sqrt{\frac{5}{2}}, -\frac{3}{2}\}$
9	$\{\sqrt{\frac{3}{2}}, \sqrt{\frac{3}{2}}, -\sqrt{\frac{5}{2}}, -\frac{3}{2}\}$
10	$\{\sqrt{\frac{3}{2}}, \sqrt{\frac{3}{2}}, \sqrt{\frac{5}{2}}, -\frac{3}{2}\}$
11	$\{-\frac{1}{\sqrt{2}}, \frac{1}{\sqrt{2}}, -\sqrt{\frac{7}{2}}, -\frac{5}{2}\}$
12	$\{-\frac{1}{\sqrt{2}}, \frac{1}{\sqrt{2}}, \sqrt{\frac{7}{2}}, -\frac{5}{2}\}$
13	$\{\frac{1}{\sqrt{2}}, -\frac{1}{\sqrt{2}}, -\sqrt{\frac{7}{2}}, -\frac{5}{2}\}$
14	$\{\frac{1}{\sqrt{2}}, -\frac{1}{\sqrt{2}}, \sqrt{\frac{7}{2}}, -\frac{5}{2}\}$

Solve::svars: Equations may not give solutions for all "solve" variables.≫
Solve::svars: Equations may not give solutions for all "solve" variables.≫
Solve::svars: Equations may not give solutions for all "solve" variables.≫
General::stop: Further output of Solve::svars will be suppressed during this calculation.≫

No.	Point of extremum	Second order diff.	Value of f	Nature
1	$\{-2, -2, 0\}$	$4\,dx^2 - 10dz^2$	12	saddle
2	$\{-2, 2, 0\}$	$-2\left(2dx^2 + 7dz^2\right)$	20	max
3	$\{0, 0, -2\}$	$-2\left(dx^2 + dxdy + dy^2\right)$	8	max
4	$\{0, 0, 2\}$	$-2\left(dx^2 + dxdy + dy^2\right)$	8	max
5	$\{2, -2, 0\}$	$-2\left(2dx^2 + 7dz^2\right)$	20	max
6	$\{2, 2, 0\}$	$4dx^2 - 10dz^2$	12	saddle
7	$\{-\sqrt{\frac{3}{2}}, -\sqrt{\frac{3}{2}}, -\sqrt{\frac{5}{2}}\}$	$4\left(dx^2 + dxdy + dy^2\right)$	$\frac{23}{4}$	min
8	$\{-\sqrt{\frac{3}{2}}, -\sqrt{\frac{3}{2}}, \sqrt{\frac{5}{2}}\}$	$4\left(dx^2 + dxdy + dy^2\right)$	$\frac{23}{4}$	min
9	$\{\sqrt{\frac{3}{2}}, \sqrt{\frac{3}{2}}, -\sqrt{\frac{5}{2}}\}$	$4\left(dx^2 + dxdy + dy^2\right)$	$\frac{23}{4}$	min
10	$\{\sqrt{\frac{3}{2}}, \sqrt{\frac{3}{2}}, \sqrt{\frac{5}{2}}\}$	$4\left(dx^2 + dxdy + dy^2\right)$	$\frac{23}{4}$	min
11	$\{-\frac{1}{\sqrt{2}}, \frac{1}{\sqrt{2}}, -\sqrt{\frac{7}{2}}\}$	$-4dxdy$	$\frac{31}{4}$	saddle
12	$\{-\frac{1}{\sqrt{2}}, \frac{1}{\sqrt{2}}, \sqrt{\frac{7}{2}}\}$	$-4dxdy$	$\frac{31}{4}$	saddle
13	$\{\frac{1}{\sqrt{2}}, -\frac{1}{\sqrt{2}}, -\sqrt{\frac{7}{2}}\}$	$-4dxdy$	$\frac{31}{4}$	saddle
14	$\{\frac{1}{\sqrt{2}}, -\frac{1}{\sqrt{2}}, \sqrt{\frac{7}{2}}\}$	$-4dxdy$	$\frac{31}{4}$	saddle

References

1. Abell, M.L., Braselton, J.P.: Differential Equations with *Mathematica*. AP Professional, Boston (1993)
2. Adamchik, V., Wagon, S.: π A 2000-year search changes direction. Math. Educ. Res. **5**(1), 11–19 (1996). www.cs.cmu.edu/~adamchik/articles/pi/pi.htm
3. Adamchik, V., Wagon, S.: A simple formula for π. Am. Math. Mon. **104**(9), 852–855 (1997)
4. Alexander, R.: Diagonally implicit Runge-Kutta methods for stiff O.D.E's. SIAM J. Numer. Anal. **14**(6), 1006–1021 (1977)
5. Backhouse, N.: Pancake functions and approximations to π. Math. Gaz. **79**, 371–374 (1995). Note 79.36
6. Bailey, D.H., Borwein, P.B., Plouffe, S.: On the rapid computation of various polylogarithmic constants. Math. Comput. **66**(218), 903–913 (1997)
7. Baruah, D.N., Berndt, B.C., Chan, H.H.: Ramanujan's series for $1/\pi$: a survey. Am. Math. Mon. **116**(7), 567–587 (2009)
8. Bellard, F.: Computation of the n'th digit of π in any base in $O(n^2)$ (1997). fabrice.bellard. free.fr/pi/
9. Borwein, J.M., Borwein, P.B.: The class three Ramanujan type series for $1/\pi$. J. Comput. Appl. Math. **45**(1–2), 281–290 (1993)
10. Borwein, J.M., Borwein, P.B., Bailey, D.H.: Ramanujan, modular equations, and approximations to π or how to compute one billion digits of π. Am. Math. Mon. **96**(3), 201–219 (1989)
11. Borwein, J.M., Skerritt, M.P.: An Introduction to Modern Mathematical Computing with *Mathematica*. Springer Undergraduate Text in Mathematics and Technology. Springer, New York (2012)
12. Brun, V.: Carl Störmer in memoriam. Acta Math. **100**(1–2), I–VII (1958)
13. Bryson, A.E., Ho, Y.C.: Applied Optimal Control: Optimization, Estimation, and Control. Halsted Press, New York (1975)
14. Burns, R.E., Singleton, L.G.: Ascent from the lunar surface. Technical report TN D-1644, NASA, George C. Marshall Space Flight Center, Huntsville (1965)
15. Cesari, L.: Optimization–Theory and Applications. Problems with Ordinary Differential Equations. Applications of Mathematics, vol. 17. Springer, New York (1983)
16. Champion, B.: General Visualization Quick Start. Wolfram Research, Champaign (2013). www.wolfram.com/training/courses/vis412.html
17. Chudnovsky, D.V., Chudnovsky, G.V.: The computation of classical constants. Proc. Natl. Acad. Sci. USA **86**(21), 8178–8182 (1989)
18. Cloitre, B.: A BBP formula for π^2 in golden base (2003). abcloitre@wanadoo.fr
19. Dennis Lawrence, J.: A Catalog of Special Plane Curves. Dover, New York (1972)

© Springer International Publishing AG 2017
M. Mureşan, *Introduction to Mathematica® with Applications*,
DOI 10.1007/978-3-319-52003-2

20. Don, E.: Mathematica. Schaum's Outlines Series. McGraw Hill, New York (2009)
21. Ebbinghaus, H.D., Peckhaus, V.: Ernst Zermelo. An Approach to His Life and Work. Springer, Berlin/Heidelberg (2007)
22. Elkies, N.D.: On $a^4 + b^4 + c^4 = d^4$. Math. Comput. **51**(184), 825–835 (1988)
23. Fay, T.H.: The butterfly curve. Am. Math. Mon. **96**(5), 442–443 (1989)
24. Ferguson, H., Gray, A., Markvorsen, S.: Costa's minimal surface via *Mathematica*. Math. Educ. Res. **5**(1), 5–10 (1966)
25. Finch, S.R.: Zermelo's navigation problem. In: Mathematical Constants II. Encyclopedia of Mathematics and Its Applications. Cambridge University Press, Cambridge (Forthcoming)
26. Floyd, R.W.: Algorithm 245: treesort. Commun. ACM **7**(12), 701 (1964)
27. Frye, R.E.: Finding $95800^4 + 217519^4 + 414560^4 = 422481^4$ on the connection machine. In: Proceedings of Supercomputing'88. Science and Applications, vol. 2, pp. 106–116 (1988)
28. Gourevitch, B., Guillera Goyanes, J.: Construction of binomial sums for π and polylogarithmic constants inspired by BBP formula. Appl. Math. E-Notes **7**, 237–246 (2007). www.math.nthu.edu.tw/~amen
29. Gradshteyn, S.G., Ryzhik, I.M.: Tables of Integrals, Series, and Products, 7th edn. Elsevier, Amsterdam (2007)
30. Guillera, J., Zudilin, W.: Ramanujan-type for $1/\pi$: the art of translation. In: Bruce, D.P., Berndt, C. (eds.) The Legacy of Srinivasa Ramanujan. Lecture Notes Series, vol. 20, pp. 181–195. Ramanujan Mathematical Society (2013). arXiv 1302.0548
31. Hartman, P.: Ordinary Differential Equations, 1st edn. Wiley, Hoboken (1964)
32. Hastings, C., Mischo, K., Morrison, M.: Hands-On Start to Wolfram *Mathematica*® and Programming with the Wolfram Language^{TM}. Wolfram Media, Champaign (2015)
33. Hazrat, R.: Mathematica®: A Problem-Centered Approach. Springer Undergraduate Mathematics Series, vol. 53. Springer, London (2010)
34. Hoare, C.A.R.: Algorithm 64: quicksort. Commun. ACM **4**(7), 321 (1961)
35. Hull, D.G.: Optimal guidance for Lunar ascent. Adv. Astronaut. Sci. **134**, 275–285 (2009). Proccedings of the AAS Space Flight Machanics Meeting, Savannach
36. Hull, D.G.: Optimal guidance for quasi-planar Lunar ascent. J. Optim. Theory Appl. **151**(2), 353–372 (2011)
37. Hull, D.G., Harris, M.W.: Optimal solutions for quasiplanar ascent over a spherical Moon. J. Guid. Control Dyn. **35**(4), 1218–1224 (2012). doi: 10.2514/1.55443
38. Knuth, D.E.: The Art of Computer Programming. Sorting and Searching. Computer Science and Information Processing, vol. 3. Addison-Wesley, Reading (1973)
39. Lander, L.J., Parkin, T.R.: Counterexample to Euler's conjecture on sums of like powers. Bull. Am. Math. Soc. **72**(6), 1079 (1966)
40. Lawden, D.F.: Analytical Methods of Optimization. Dover Books on Mathematics. Dover, Mineola (2006)
41. Lucas, S.K.: Integral proofs that $355/113 > \pi$. Gaz. Aust. Math. Soc. **32**(4), 263–266 (2005)
42. Lucas, S.K.: Integral approximations to π with nonnegative integrands (2007). carma.newcastle.edu.au/jon/Preprints/Papers/By%20Others/more-pi.pdf
43. Mangano, S.: Mathematica Cookbook. O'Reilly, Sebastopol (2010)
44. Manià, B.: Sopra un problema di navigatione di Zermelo. Math. Ann. **113**(1), 584–589 (1937)
45. McShane, E.J.: A navigation problem in the calculus of variations. Am. J. Math. **59**(2), 327–334 (1937)
46. Miele, A.: Flight Mechanics. Theory of Flight Path. Addison-Wesley Series in the Engineering Sciences Space Science and Technology, vol. 1. Addison-Wesley, Reading (1962)
47. Mureşan, M.: Classical Analysis by Mathematica (Forthcoming)
48. Mureşan, M.: A Primer on the Calculus of Variations and Optimal Control. Trajectories Optimization (Forthcoming)
49. Mureşan, M.: On a Runge-Kutta type method. Rev. Anal. Numér. Théorie Approx. **16**(2), 141–147 (1987)
50. Mureşan, M.: Some computing results of a Runge-Kutta type method. Seminar on Mathematical Analysis Nr. 7, Univ. Babeş-Bolyai, Cluj-Napoca, pp. 101–114 (1987)

51. Mureşan, M.: A semi-explicit Runge-Kutta method. Seminar on Differential Equations Nr. 8, Univ. Babeş-Bolyai, Cluj-Napoca, pp. 65–70 (1988)
52. Mureşan, M.: Qualitative Properties of Differential Equations and Inclusions. Ph.D. thesis, Babeş-Bolyai University, Cluj-Napoca (1996)
53. Mureşan, M.: A Concrete Approach to Classical Analysis. CMS Books in Mathematics. Springer, New York (2009)
54. Mureşan, M.: Instructor Solution Manual, for A Concrete Approach to Classical Analysis. Springer, New York (2012). www.springer.com/mathematics/analysis/book/978-0-387-78932-3?changeHeader
55. Mureşan, M.: Soft landing on Moon with *Mathematica*. Math.® J. **14** (2012). doi: dx.doi.org/doi:10.3888/tmj.14–16
56. Mureşan, M.: On Zermelo's navigation problem with *Mathematica*. J. Appl. Funct. Anal. **9**(3–4), 349–355 (2014)
57. Mureşan, M.: On the maximal orbit transfer problem. Math.® J. **17** (2015). doi: dx.doi.org/10.3888/tmj.17–4
58. Palais, R.S.: The visualization of mathematics: towards a mathematical exploratorium. Not. Am. Math. Soc. **46**(6), 647–658 (1999)
59. Rao, K.S.: Ramanujan and important formulas. In: Nagarajan, K.R., Soundararajan, T. (eds.) Srinivasa Ramanujan: 1887–1920: A Tribute, pp. 32–41. MacMillan India, Madras (1988)
60. Robinson, E.A., Jr.: Man Ray's human equations. Not. Am. Math. Soc. **62**(10), 1192–1198 (2015)
61. Sedgewick, R.: Implementing Quicksort programs. Commun. ACM **21**(10), 847–857 (1978)
62. Strogatz, S.H.: Nonlinear Dynamics and Chaos: with Applications to Physics, Biology, Chemistry, and Engineering. Perseus, New York (1994)
63. Takano, K.: Pi no arctangent relation wo motomete [finding the arctangent relation of π]. Bit **15**(4), 83–91 (1983)
64. Torrence, B., Torrence, E.: The Student's Introduction to *Mathematica*®. A Handbook for Precalculus, Calculus, and Linear Algebra, 2nd edn. Cambridge University Press, Cambridge, UK (2009)
65. Trott, M.: The *Mathematica* GuideBook for Graphics. Springer, New York (2004)
66. Trott, M.: The *Mathematica* GuideBook for Programming. Springer, New York (2004)
67. Trott, M.: The *Mathematica* GuideBook for Numerics. Springer, New York (2006)
68. Trott, M.: The *Mathematica* GuideBook for Symbolics. Springer, New York (2006)
69. Weisstein, W.E.: BBP-Type formula. Technical report, *MathWorld*-A Wolfram Web Resources. mathworld.wolfram.com/BBP-TypeFormula.html
70. Weisstein, W.E.: Machbin-Like formulas. Technical report, *MathWorld*-A Wolfram Web Resources. mathworld.wolfram.com/Machin-LikeFormulas.html
71. Weisstein, W.E.: Pi formulas. Technical report, *MathWorld*-A Wolfram Web Resources. mathworld.wolfram.com/PiFormulas.html
72. Wolfram, S.: The *Mathematica*® Book, 5th edn. Wolfram Media, Champaign (2003)
73. Wolfram, S.: An Elementary Introduction to the Wolfram Language. Wolfram Media, Champaign (2015)
74. Wolfram, S.: Differential Equation Solving with DSOLVE. Wolfram Research, Champaign (2008). Wolfram Mathematica® Tutorial Collection. htpps:/reference.wolfram.com/language/tutorial/DSolveOverview.html
75. Wolfram, S.: Advanced numerical differential equation solving in *Mathematica*. In: Wolfram Mathematica® Tutorial Collection. Wolfram Research, Champaign (2008). htpps:/www.scrib.com/doc/122203558/Advanced-Numerical-Differential-Equation-Solving-in-Mathematica
76. Wolfram, S.: http://numbers.computation.free.fr/Constants/Pi/piclassic.html
77. Wolfram, S.: http://numbers.computation.free.fr/Constants/Pi/piramanujan.html
78. Zermelo, E.: Über die Navigation in der Luft als Problem der Variationsrechnung. Jahresbericht der deutschen Mathematiker – Vereinigung, Angelegenheiten **39**, 44–48 (1930)
79. Zermelo, E.: Über das Navigationproblem bei ruhender oder veränderlicher Windverteilung. Z. Angew. Math. Mech. **11**(2), 114–124 (1931)

Index

© Springer International Publishing AG 2017
M. Mureşan, *Introduction to Mathematica® with Applications*,
DOI 10.1007/978-3-319-52003-2

Printed in the United States
By Bookmasters